Biotechnology and Plant Breeding

Applications and Approaches for Developing Improved Cultivars

Biotechnology and Plant Breeding

Applications and Approaches for Developing Improved Cultivars

Aluízio Borém
Federal University of Viçosa, Viçosa, MG, Brazil
Roberto Fritsche-Neto
University of São Paulo / ESALQ, Piracicaba, SP, Brazil

AMSTERDAM • BOSTON • HEIDELBERG • LONDON
NEW YORK • OXFORD • PARIS • SAN DIEGO
SAN FRANCISCO • SINGAPORE • SYDNEY • TOKYO
Academic Press is an imprint of Elsevier

Academic Press is an imprint of Elsevier
32 Jamestown Road, London NW1 7BY, UK
225 Wyman Street, Waltham, MA 02451, USA
525 B Street, Suite 1800, San Diego, CA 92101-4495, USA

First edition 2014

Notice
No responsibility is assumed by the publisher for any injury and/or damage to
persons or property as a matter of products liability, negligence or otherwise,
or from any use or operation of any methods, products, instructions or ideas
contained in the material herein.

Because of rapid advances in the medical sciences, in particular, independent
verification of diagnoses and drug dosages should be made.

British Library Cataloguing-in-Publication Data
A catalogue record for this book is available from the British Library

Library of Congress Cataloging-in-Publication Data
A catalog record for this book is available from the Library of Congress

ISBN: 978-0-12-418672-9

For information on all Academic Press publications
visit our website at elsevierdirect.com

Typeset by Scientific Publishing Services (P) LTD
www.sps.co.in

Working together
to grow libraries in
developing countries

www.elsevier.com • www.bookaid.org

Contents

Contributors

Camila Ferreira Azevedo, Federal University of Viçosa, Viçosa, Brazil

Marcos Fernando Basso, Federal University of Viçosa, Viçosa, Brazil

Leonardo Lopes Bhering, Federal University of Viçosa, Viçosa, Brazil

Aluízio Borém, Federal University of Viçosa, Viçosa, Brazil

Eveline Teixeira Caixeta, Embrapa Coffee, Brasilia, Brazil

Gloria Patricia Castillo Urquiza, Federal University of Viçosa, Viçosa, Brazil

Cosme Damião Cruz, Federal University of Viçosa, Viçosa, Brazil

Fábio Nascimento da Silva, Federal University of Viçosa, Viçosa, Brazil

Marcos Deon Vilela de Resende, Brazilian Agricultural Research Agency/Federal University of Viçosa, Viçosa, MG, Brazil

Valdir Diola, Federal Rural University of Rio de Janeiro, Seropédica, RJ, Brazil (*in memorium*)

Luis Felipe Ventorim Ferrão, Federal University of Viçosa, Viçosa, Brazil

Fabyano Fonseca e Silva, Federal University of Viçosa, Viçosa, Brazil

Roberto Fritsche-Neto, University of São Paulo, São Paulo, Brazil

Deoclecio Domingos Garbuglio, Agronomic Institute of Paraná, Londrina, Brazil

Eunize Maciel-Zambolim, Federal University of Viçosa, Viçosa, Brazil

Moacir Pasqual, Federal University of Lavras, Brazil

Márcio Fernando R. Resende Júnior, University of Florida, Gainesville, FL, USA

Filipe Almendagna Rodrigues, Federal University of Lavras, Brazil

Joyce Dória Rodrigues Soares, Federal University of Lavras, Brazil

Caio Césio Salgado, Federal University of Viçosa, Viçosa, Brazil

Natália Arruda Sanglard, Federal Rural University of Rio de Janeiro, Seropédica-RJ, Brazil

Laércio Zambolim, Federal University of Viçosa, Viçosa, Brazil

Francisco Murilo Zerbini, Federal University of Viçosa, Viçosa, Brazil

Plant breeding is a relatively new science, little more than a century old. Using plant breeding, scientists have developed improved cultivars, especially using methods developed in the 1900s. However, the challenges have become daunting for agriculture to supply food, feed, fiber, and biofuel for an increasingly resource-hungry world. Therefore, plant breeding must evolve and make use of new technologies to become more efficient and accurate at developing new and improved cultivars. It is not surprising that biotechnology has been gradually incorporated into plant breeding over the last few decades. Evidence for this is provided by the large numbers of genetically modified cultivars being cultivated around the world. The main biotechnology tool used by breeders today is recombinant DNA technology; however, many other biotechnology tools are now available and are being successfully used in cultivar development such as double-haploids, molecular markers, genome-wide selection, among others.

The book *Biotechnology and Plant Breeding: Applications and Approaches for Developing Improved Cultivars* offers outstanding opportunities to contribute to the well-being of mankind. This contribution to mankind comes about when improved varieties of food, feed, fiber, and fuel crops are developed that increase productivity and security. Such varieties have become available through traditional breeding and biotechnology and are being grown by large as well as small farmers. Ultimately the improved performance benefits society as a whole. Furthermore, opportunities exist to increase these contributions in the future through the application of biotechnology in plant-breeding programs.

This book was written by a select group of knowledgeable scientists, and was designed to be used as a textbook and a reference book for scientists and students around the world interested in the new biotechnologies applied to plant breeding.

Aluízio Borém and Roberto Fritsche-Neto
Editors

Chapter 1

Plant Breeding and Biotechnological Advances

Aluízio Borém,[a] Valdir Diola,[b] and Roberto Fritsche-Neto[c]

[a]*Federal University of Viçosa, Viçosa, Brazil,* [b]*Federal Rural University of Rio de Janeiro, Rio de Janeiro, Brazil,* [c]*University of São Paulo, São Paulo, Brazil*

INTRODUCTION

Currently, the global population comprises more than 7 billion people, and the global population clock is currently recording continuous growth (http://www.apolo11.com/populacao.php). Such growth will continue until approximately 2050, the year during which population growth is expected to plateau at the staggering number of 9.1 billion people, according to United Nations (UN) predictions. It is notable that thousands of years were needed to increase the global population to the initial 2 billion people, yet another 2 billion will be added to the planet in the next 25 years (Figure 1.1).

Additionally, people are living longer and migrating from rural areas to cities. Furthermore, the population's purchasing power and land competition for grain and renewable energy production are increasing (Beddington, 2010). Therefore, all current food production systems must either double in productivity until 2050 (Clay, 2011) or risk failing to meet the growing demand for food, thus materializing Malthus's predictions of mass starvation, which were made approximately 200 years ago.

The challenges of feeding the world are tremendous and have led scientists to seek more efficient food production methods. In that context, many innovations are being incorporated into food production in order to meet that growing demand, including precision farming, micronutrient use, protected cultivation and integrated crop management, among others. Among other innovations, cultivar development from plant genetic breeding is considered one of the most important and has been responsible for more than half of the increases in crop yields over the last century.

Many definitions of plant breeding have been introduced by different authors, including evolution directed by the will of man (Vavilov, 1935), the genetic adjustment of plants to the service of man (Frankel, 1958), an exercise in exploring the genetic systems of plants (Williams, 1964), the art and science

Biotechnology Applied to Plant Breeding. http://dx.doi.org/10.1016/B978-0-12-418672-9.00001-5

1

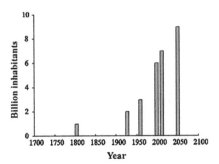

FIGURE 1.1 World population growth throughout the years.

of improving the heredity of plants for the benefit of mankind (Allard, 1971), the exploration of the genetic potential of plants (Stoskopf et al., 1993), and the science, art, and business of improving plants for human benefit (Bernardo, 2010).

Undoubtedly, plant breeding enables agriculture to sustainably provide foods, fibers, and bioenergy to society. For example, breeding develops forage and grains for animal feed to support meat, milk, and egg production. Agro-bioenergy activities require the development of more efficient cultivars for power generation through combustion, ethanol, and biodiesel. In the future, breeding will also enable a drastic shift in the agriculture paradigm towards the production of other materials, including drugs, biopolymers, and chemicals.

EVOLUTION OF GENETICS AND PLANT BREEDING

Since the beginning of agriculture in approximately 10,000 BC, people have consciously or unconsciously selected plants with superior characteristics for the cultivation of future generations. However, there is controversy regarding the time when breeding became a science. Some believe that this occurred after Mendel's findings, while others argue that it occurred even before the "era of genetics."

One of the most important contributions to plant breeding was artificial plant hybridization, which permitted the gathering of advantageous characteristics into a single genotype. Consequently, some dates and events indicate the beginning of this new science, such as August 25, 1964, when R.J. Camerarius published the article "De sex plantarum epístola," or even 1717, when Thomas Fairchild created the first hybrid plant in England. In addition to those events, J.G. Kolreuter conducted the first scientific experiment on plant hybridization in 1760.

During the nineteenth century, plant breeding had already begun in France, as Louis Vilmorin had developed wheat and sugar beet varieties with progeny tests. However, the monk Gregor Mendel from Brno, Czech Republic, unveiled the secrets of heredity and thus ushered in the "era of genetics," the fundamental science of plant breeding, at the end of that century.

By placing a few more pieces into the puzzle of this new science, scientists in the first half of the twentieth century knew that something within cells was responsible for heritability. That hypothesis started a process of hypothesis generation and discovery, thus further enabling progress and knowledge accumulation in the field to continue apace. For example, the DNA double helix structure was elucidated in 1953 (Table 1.1). Twenty years later, in 1973, the discovery of restriction enzymes opened the doors of molecular biology to scientists. The first transgenic plant, wherein a bacterial gene was stably inserted into a plant genome, was created in 1983.

At that time, futuristic predictions about biotechnology contributions were reported in the media by both laymen and scientists, and these created great expectations for their applications. This euphoria was a keynote in the scientific community. Many large and small companies were created in response to the prevailing enthusiasm at the time, although most later went bankrupt (Borém and Miranda, 2013). The failures occurred because most biotechnology predictions did not materialize according to the initially predicted schedule, and thus skepticism led many of those entrepreneurs to face reality and the investors to relocate their resources.

Currently, the results of many earlier predictions have materialized (Table 1.1), which has led to a consensus that the benefits of biotechnology will have greater impacts on breeding programs each year. Consequently, new companies are being established under the prospects of a highly promising market.

THE IMPACTS OF ADVANCES DURING THE TWENTIETH CENTURY

The twentieth century was marked by great discoveries that profoundly affected plant-breeding methods, starting with the rediscovery of Mendel's laws in 1900. In 1909, Shull wrote the first study on the use of heterosis in maize. The 1920s were marked by the development of classical breeding methods. In the 1930s, euphoria resulted from the discovery of mutagenesis and the use of statistical methods. There were great advances in quantitative genetics in the 1940s, in physiology in the 1950s, in biochemistry in the 1960s, in tissue culture in the 1970s, in molecular biology in the 1980s (Borém and Miranda, 2013), and in genetic transformation in the 1990s.

Based on experience gained during 45 years of intense and effective plant breeding in both the private and academic sectors, Donald Duvick (1986) reported that the greatest advances he witnessed during that period were the adoption of mechanization in experimentation and the use of computers in breeding programs; the practice of two or more generations per year, which reduced the time needed to create new cultivars; and increases in communication speed between breeders worldwide, which transformed breeding methods. Scientific knowledge-based breeding has enabled agricultural production to meet the global demand for food (Wolf, 1986).

TABLE 1.1 Chronology of the Historical Facts Related to Key Advances in Genetics and Biotechnology That Are Relevant to Plant Breeding

Year	Historic Landmark
1744 to 1829	Lamarck describes the hypothesis of the hereditary transmission of acquired characters. Periodic observations that disregarded the effects of selection, adaptation, and mortality induced the erroneous conclusion that anatomical characteristics change according to the environmental requirements.
1809 to 1882	Charles Darwin writes the theory of natural selection, which was described in the book *The Origin of Species*, according to observations collected on the Galapagos islands. The species that best adapted to their environment were selected to survive and produce further offspring.
1865	Gregor Mendel establishes and applies the first statistical methods to pea seed breeding, thus marking the beginning of the "genetic era."
1876	The first interspecific cross between wheat and rye to yield triticale.
1910	Thomas Morgan shows that genetic factors (genes) are located in chromosomes while studying the effects of genetic recombination in *Drosophila melanogaster*.
1923	Karl Sax reports the study of quantitative trait loci (QTL) based on the pigmentation and coloration traits of beans.
1928	Griffith finds that the same line of infectious bacteria could be either virulent or not in the presence or absence of genetic factors, thereby beginning the clarification of the chemical nature of DNA.
1941	George Beadle and Edward Tatum show that a gene produces a protein.
1944	Barbara McClintock explains the process of genetic recombination by studying satellite chromosomes and the genetic linkage regarding the linkage groups 8 and 9 in maize.
1944	Avery, MacLeod, and McCarty continue Griffith's experiment and find that DNA is the material responsible for heredity by using nucleases and proteases.
1953	James Watson and Francis Crick propose the double helical structure of the DNA molecule by using X-ray diffraction.
1957	Hunter and Markert develop biochemical markers based on enzyme expression (isoenzymes), with applicability in genotypic selection.
1969	Herbert Boyer discovers restriction enzymes and thus introduces new prospects for DNA fingerprinting and the cloning of specific regions.

(Continued)

TABLE 1.1 Continued

Year	Historic Landmark
1972	Initial recombinant DNA technology is introduced with the first cloning of a DNA fragment.
1973	Stanley Cohen and Herbert Boyer conduct the first genetic engineering experiment on a microorganism, *Escherichia coli*. The result was considered the first genetically modified organism (GMO).
1977	Maxam and Gilbert develop DNA sequencing by chemical degradation.
1980	Botstein et al. develop the RFLP (Restriction Fragment Length Polymorphism) method for the wide use of genotypic selection.
1975	Sanger develops sequencing with an enzymatic method; in 1984, the method is improved and the first automatic sequencers are built in the 1980s.
1983	The first transgenic plant, a variety of tobacco into which a group of Belgian scientists introduced kanamycin antibiotic-resistance genes, is created.
1985	Genentech becomes the first biotech company to launch its own biopharmaceutical, human insulin produced from an *E. coli* culture that was transformed with the functional human gene.
1985	Production of the first plant with the Lepidoptera resistance gene.
1986	The first experimental transgenic plant field in Gant, Belgium.
1987	The first plant tolerant to an herbicide, glyphosate, is created.
1987	Mullis and Faloona identify the thermostable Taq DNA polymerase enzyme and thus enable automated PCR.
1988	The first transgenic cereal, the Bt grain, is created.
1990	Creation of the new National Center for Biotechnology Information (NCBI) sequence alignment tools (Basic Local Alignment Search Tool (BLAST); www.ncbi.gov).
1991	Rafalski et al. (1991) develop RAPD (Random Amplified Polymorphism DNA), the first PCR genotyping method.
1994	The world's first license for the commercial planting of a GMO, the Flavr Savr tomato, is released.
1997	First plant with a human gene, a human protein C-producing tobacco, is produced.
2000	The total area planted with transgenic crops worldwide is 44.2 million hectares. The United States, Argentina, Canada, Australia, South Africa, Romania, Bulgaria, France, Spain, Uruguay, and Ukraine are GMO producers.

(Continued)

TABLE 1.1 Continued

Year	Historic Landmark
2000	First complete sequencing of a prokaryotic organism, the bacteria *E. coli*, is reported.
2003	The sequence of the first eukaryote genome, human, is published by two large independent research groups in the United States.
2004	Eighteen countries are planting GM varieties on a total of 67.7 million hectares.
2005 to 2009	Large-scale sequencing (Next Generation Sequencing (NGS)) is used as a tool to unveil complete genomes in a short period of time.
2006 to 2011	Second- and third-generation genetically modified (GM) plants with specific characteristics are developed in well-studied species.
2010	The area planted with GM crops worldwide exceeds 180 million hectares, and Brazil is the second largest producer of genetically modified crops.
2011	Large-scale second- and third-generation sequencers permit eukaryotic genomes to be sequenced in just a few hours.
2012	Technologies are used to control the temporal and spatial expression of genes used in genetic transformation and auxiliary gene exclusion.
2013...	Widespread usage of large-scale sequencing, macro- and micro-synteny, associative mapping, molecular markers for genome-wide selection, QTL cloning, large-scale phenotyping, expanded use of "omics," specific GMOs, multiple phenotypic attributes, and the intense use of bioinformatics.

The Green Revolution was certainly an example of the economic and social impacts that plant breeding could make worldwide. The introduction of dwarfing genes into wheat and subsequently into other cereal crops, including rice, enabled significant increases in the adoption and yield of those species. Dwarfing genes also enabled increased nitrogen fertilizer use and thus generated the use of technological packages even in third world countries, with subsequent significant increases in global food production (Borlaug, 1968, 1969).

Plant breeders have relied on the help of some valuable tools to reach their goals. Two of the main evolutionary factors, recombination and selection, have been used intensively by breeders through the application of refined methods that were developed during the first half of the twentieth century. In recent years, a highly promising new tool, molecular biology, has emerged.

Most breeders believed that the main applications of molecular markers would be monogenic factor introgression, marker-assisted breeding, quantitative trait locus (QTL), and transgene transfer and parent selection (Lee, 1995). According to Lee (1995), the main limitations to the uses of molecular markers in cultivar development would be proper and large-scale phenotyping, technology costs, genotype-marker interactions, and operational difficulties.

At the turn of the 21st century, genetic research studies in universities worldwide were predominantly directed towards plant molecular biology and genetic transformation. This resulted from the high level of specificity in research lines in genetics and difficulties in breeding some traits through hybridization or, even, from the inexistence of genetic resources. Another reason for the use of genetic transformation relates to the study of the expression of genes of commercial interest. Consequently, the number of universities that noticeably allocate human and financial resources to this field has rapidly increased. Thus, biotechnology is considered a valuable tool in breeding programs and is increasingly being used, enabling breeders and biotechnologists to understand each other more easily than in the early days of biotechnology.

ADVANCES AND EXPECTED BENEFITS OF BIOTECHNOLOGY

Based on the facts mentioned earlier in this chapter, a new Green Revolution may be necessary to increase worldwide food production and meet future demands. Thus, the question arises: Can biotechnology bring plant breeding to a new Green Revolution? Evidence of this is already believed to exist.

One piece of evidence is that the number of transgenic cultivars released into the market in recent years has increased substantially. Herbicide-tolerant cultivars are prevalent for most commodity species, followed by those with insect resistance and others. Biotechnology contributions to agriculture are already felt in many countries wherein transgenic cultivars occupy large areas of arable land with different species, including the United States, Brazil, Argentina, Canada, and several others. Eventually, breeding might achieve yield plateaus due to the restrictions imposed by the pyramiding (Milach and Cruz, 1997) of genes available for biotechnology or those existent in the germplasm, leading to what is termed gene arrest. However, biotechnology will create currently unimagined prospects that will enable breeders to overcome the current limitations. Transgenic plants are only part of the contributions that biotechnology has promised to plant breeding.

Tanksley and McCouch (1997) highlighted the importance of the use of genetic resources in genebanks and those often found in wild species for such contributions to be achieved. It is possible to access the genetic variability in DNA with molecular markers, which will revolutionize how genetic variability will be explored in future breeding programs.

IMPROVEMENT OF TOOLS IN THE THIRD MILLENNIUM

Recent advances in genomics have enabled previously unreachable goals, including the sequencing and unraveling of genetic information contained in the genome of an organism, to be achieved in a relatively short period. The following advances stand out among those technologies: advances in tissue culture methods, the development of molecular markers that cover the entire genome and are linked to desirable traits, the design of genetic maps for all chromosomes of an organism, the development and automation of large-scale complementary DNA (cDNA) and genomic DNA sequencing methods, and the processing and refinement of genetic algorithms to analyze large amounts of data (bioinformatics).

Tissue Culture

The totipotency and ability of an organ or tissue to regenerate from a cell was recognized from observations of plant regeneration from injured tissues. Recognition of this biological principle is credited to the German plant physiologist Haberlandt, who ruled in 1902 that each plant cell had the genetic potential to develop into a complete organism. Haberlandt predicted that tissues, organs, and cells could be maintained indefinitely in culture. Tissue culture thus relates to the development of tissues and/or *in vitro* cell systems; in other words, those separated from the material source organism and maintained in a culture medium that contains carbohydrates, vitamins, hormones, minerals, and other nutrients essential to cultured tissue growth.

Plant tissue culture is a key method that has been used in conventional genetic breeding to broaden genetic variability through somaclonal variation, reduce the time for new cultivar development through anther cultures (Figure 1.2), introduce genes of agronomic interest through embryo rescue, and perform viral cleaning or

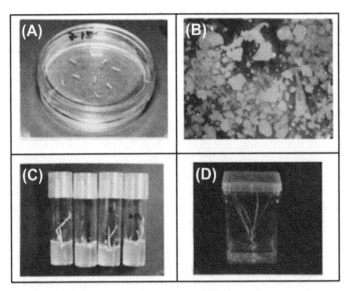

FIGURE 1.2   **Somatic embryogenesis in barley.** (A) Anthers in nutrient medium, (B) embryoid, (C) seedlings, and (D) completely regenerated plant.

indexing through meristem culture, to provide a few examples. Plant tissue culture, an essential biotechnological tool, has been widely used to develop genetically modified varieties through the regeneration of full plants from transformed cells. Genetic transformation alone also enables increases in the number of available genes in a species' gene pool. Additionally, tissue culture has been adopted as a practice of genetic resource conservation in various breeding programs.

Another tissue culture application is the production of homozygous lines through dihaploids, which could reduce the time required to develop new cultivars (Belicuas et al., 2007). This method, albeit not recent, needs to be optimized to increase efficacy. However, it is already used commercially in several agronomic species, including wheat, barley, and maize (Murovec and Bohanec, 2012).

Genetic Engineering

Genetically modified or transgenic plants, as they are popularly known, are those in which the genomes have been altered by introducing exogenous DNA through different transformation methods. Exogenous DNA can be derived from other specimens of the same species or from an altogether different species and can even be artificial (e.g., synthesized in the laboratory). The term genetically modified organism (GMO) is also frequently and generically used to indicate any individual that has been genetically manipulated through recombinant DNA methods.

The biggest bottleneck in transgenic plant creation was the availability of cloned genes. The structural and functional analysis and identification of bacterial and viral genes were adequate, and thus the studies of these genes progressed more rapidly in those organisms than in higher plants. Consequently, the transfer of viral or bacterial genes into higher plants was highlighted in initial gene transfer experiments.

The first transgenic product commercialized worldwide was the Flavr-Savr tomato, which was launched in 1996 in the United States (Table 1.1) and developed through recombinant DNA methods by the Calgene Company with the intent to slow post-harvest ripening. Today, several other products are currently available in the market, including the cotton cultivar "Ingard," which was launched in 1996. "Ingard" carries the Bt gene from *Bacillus thuringienses*, which confers resistance to larvae. The "Roundup Ready" soybean, glyphosate herbicide-tolerant maize hybrids that carry the Bt gene, and insect-resistant and herbicide-tolerant tomato cultivars are the most successful examples of transgenics. Other examples are virus-resistant bean cultivars and potato clones and canola cultivars with improved oil quality (Kubicek, 1997).

During that period and currently, the traits of agronomic interest were highlighted in most cases since research studies from the beginning of this millennium focused on functional analyses of higher plant genes, leading to changes in gene traits and production. The expectation is that throughout the years, transgenic cultivars will feature increasingly complex events, higher added values, and uses from agriculture to the pharmaceutical industry (Figure 1.3). These

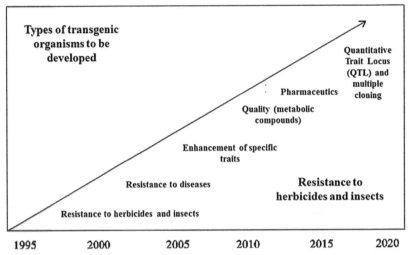

FIGURE 1.3   Expected launch of new transgenic cultivars. (*Adapted from Wenzel, 2006*)

advances will change the agriculture paradigm from a source of food, fiber, and energy to a source of chemical and pharmaceutical products.

Molecular Markers and Genetic Maps

The use of molecular markers or polymorphic DNA sequences as chromosomal references is the basic technology behind the designing of genetic maps for various species (Borém and Caixeta, 2009). With the exception of a few model organisms, the vast majority of species had no genetic map information until the late 1980s. Accordingly, there has been a great revolution and expansion of information about linkage groups in the genomes of various organisms in a short period of time. It should be noted that several branches of genetics, some of which lack in-scale experimental data such as population genetics, have also benefited from the large amounts of data generated by the use of molecular markers.

The first DNA marker-based genetic maps of plants were designed between 1986 and 1988 for maize, tomato, and rice (Bernatzky and Tanksley, 1986; Helentjaris et al., 1986). Since then, genetic maps have been designed for a great number of plant species. In general, map data have been the basis for chromosome walking towards genes of interest or chromosome landing (Tanksley et al., 1995).

Some loci that control qualitative traits were mapped in the early 1990s, including the Pto locus in tomato, which confers resistance to *Pseudomonas syringae* pv. tomato. Similar examples have been found in other plant species such as rice, wherein the sd1 gene (encodes an enzyme in the metabolic pathway of gibberellin, a plant growth hormone) confers the "semi-dwarf" phenotype upon plants. The sd1 gene was used to develop smaller modern rice cultivars that respond better to nitrogen fertilizers and do not develop eyespots. Therefore, the rice plant will produce more grains when fertilized, which will thus enable a mechanical harvest and increase productivity. For example, the "semi-dwarf" rice cultivar IR-8, developed in the 1960s, was known as "miracle rice" because it prevented an enormous expected lack-of-food crisis, especially in Asian countries. The yield of IR-8, resulting from the introduction of the sd1 gene, was two- to three-fold higher than other common varieties that existed before the Green Revolution.

Restriction fragment length polymorphisms (RFLP), random amplifications of polymorphic DNA (RAPD), amplified fragment length polymorphisms (AFLP), microsatellites or simple sequence repeats (SSR), single nucleotide polymorphisms (SNP) and derived markers, sequenced characterized amplified regions (SCAR), and characterized amplified polymorphisms sequenced (CAPS) stand out among the various types of molecular markers. The finding that simple-sequence repeats also occur in expressed sequence tags (EST) provided an opportunity to simply and directly produce microsatellite markers through online searches of EST databases. For example, Scott et al. (2000) used this methodology to generate 10 EST-derived SSR markers from a grape EST database that contained 5000 ESTs.

Genomics

The study of genomics began with the sequencing of genetic material from organisms. During its short existence, genomics has significantly advanced through robotics and large-scale gene expression analysis. Some of the genomic applications in breeding programs include molecular diagnosis (such as gene transference tests), target allele confirmation in breeding populations, genetic background testing, genomic identification, genetic identity tests for cultivar protection, varietal purity control, genetic variability organization in germplasm collections, and molecular marker-assisted selection. Technology enables phylogenetic gene synteny studies, molecular marker development, physical mapping and sequence annotation. It is noteworthy that all the genomics knowledge must be stored in databases, some of which are public domain databases, while others are restricted-access and private domain databases. Currently, the largest open-access public database is administered by the NCBI.

Genomic breeding aims to integrate conventional breeding methodologies and techniques with molecular biology technologies and approaches. This includes the use of genetic information derived from linkage maps that were designed with molecular markers in backcross programs. It also includes the pyramiding of resistance genes to different diseases through marker-assisted selection or even the use of selection indexing when defining specimens to be recombined in recurrent selection programs, based on field-collected phenotypic information and laboratory-collected DNA information (Lande and Thompson, 1990). Thus, the use of molecular markers plays a key role in genomic breeding, especially in accurate data collection, with costs and rapid diagnoses that are compatible with usage on the scale required by breeding programs.

Other "Omics"

In addition to genomics, other new sciences are being developed that will have great applicability to plant breeding. The transcriptome methods are promising because these generate a greater set of functional genetic material data. Many methods applied to the study of plant gene expression can be found in the literature, especially those based on cDNA synthesis from an mRNA strand, including Northern blotting, cDNA-AFLP, real-time PCR, subtractive hybridization, full-length libraries, differential displays, and differential expression. Transcriptome-based methods are the most powerful methods applied to gene expression studies. Currently, RNAseq is one of the most important methods because it generates a large amount of data in a short period of time at a relatively low cost.

Proteomic methods have identified several proteins and studied the processes of protein–protein, protein–DNA and protein–RNA interactions, and protein metabolism. These studies and post-transcriptional and/or post-translational regulation have aroused great interest in molecular biology and have made it an immediate tool for applications.

Metabolomics has introduced a new perspective on compounds that were previously considered to be waste, but could possibly be improved and availed by biotechnology. Soon, genetically modified plants may become biofactories.

COMBINED CLASSIC BREEDING AND BIOTECHNOLOGY

Plant breeding requires persistence because 8 to 12 years are usually required to create a new cultivar, due to the processes involved, including crosses to value for cultivation and use (VCU) testing. Biotechnology tools at least enable reductions in the selection period because unlike classical breeding, which lacks phenotyping in all generations, selection can be performed by genotyping. The integration of classical breeding methods with biotechnology tools is described in Figure 1.4. Even with the new tools, genotyping methods must initially be adjusted to be consistent with phenotyping, a relatively laborious and costly step (Cobb et al., 2013). It is worth noting that biotechnology does not replace classical breeding methods. Of particular interest is the combination of both fields towards increased selection accuracy and decreased intervals between selection generations.

Most plants have 20,000 to 60,000 genes (Arabidopsis, 25,000 and rice, 46,000; http://www.eugenes.org) and the recombination or combination of the enormous amounts of alleles through breeding is a key process in new cultivar development both now and in the future (Sharma et al., 2002). Naturally,

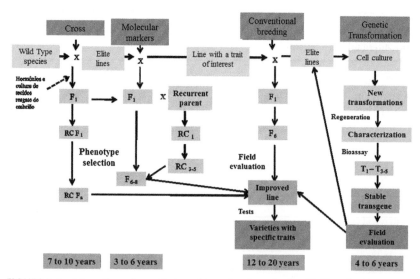

FIGURE 1.4 A breeding scheme that combines classic methods with biotechnology. Strains derived from genetic transformations can be used as donor parents in conventional breeding programs. The lines must undergo selection and homozygote generation to establish genetic stability, a process that may be more easily accomplished with molecular markers, which shorten the time period required to develop new cultivars. (*Adapted from Sharma et al., 2002*)

the inclusion of new combinations of high production performance genotypes through modern technology is a way to reduce the costs of new cultivar development because the varieties used have desirable traits and genetic stability. Increased knowledge about genetic control and the inheritance of traits of interest at the molecular level will assist in parent selection or even enable the most efficient selection of new genotypic combinations (Rommens and Kishore, 2000). It should be noted that, despite the range of new technologies, the genetic process that is documented worldwide through annual new cultivars records has primarily resulted from classical breeding; this process is expected to continue for a long time and might gradually be replaced by a biotechnology-integrated method. The extensive list of biotechnology-affiliated cultivars released into the market proves this. Currently, recombinant DNA is the main biotechnological method used to generate transgenic varieties. However, there are several methods that could be integrated into plant breeding (Table 1.2).

PERSPECTIVES

Plant breeding is an art, science, and business and is little more than a century old. Breeders have developed agronomically superior cultivars by using methods that were mainly developed in the last century. Plant breeding must evolve and incorporate new knowledge because the food production challenges for

TABLE 1.2 Biotechnology Applications for Improved Main Crop Productivity and Quality

Crop	Traits for Breeding	TC/H	MAS	Transg
Rice	Drought and salinity tolerance		X	X
	Disease resistance	X	X	X
	Grain nutritional traits and quality		X	X
	Flood resistance	.	X	
Wheat	Productivity, quality, and adaptation		X	X
	Fungi and disease resistance		X	X
Maize	Flood resistance	X	X	X
	Productivity	X	X	X
	Salinity and toxic elements	X		X
	Grain quality		X	X
Sorghum	Productivity and quality	X	X	X
	Disease resistance		X	X
	Salinity and drought tolerance	X	X	X
Soybean	Disease resistance and insect tolerance		X	X
	Herbicide resistance			X
	Oil productivity and quality		X	X
	Nutritional qualities and other compounds		X	X

(Continued)

TABLE 1.2 Continued

Crop	Traits for Breeding	TC/H	MAS	Transg
Cotton	Productivity and fiber quality		X	X
	Disease resistance and insect tolerance		X	X
	Abiotic stress tolerance (e.g., drought, salinity)	X	X	X
Eucalyptus	Fiber quality		X	X
	Productivity	X	X	X
	Essential oils	X	X	X
Sugarcane	Productivity	X	X	X
	Fiber quality		X	X
	Secondary compounds	X	X	X
	Insect, disease, and herbicide tolerance		X	X
	Abiotic stress tolerance (e.g., drought, salinity)	X	X	X
Vegetables	Nutritional qualities	X	X	X
	Pharmacological compounds			X
	Insect and disease tolerance (fungi and viruses)		X	X
	Abiotic stress tolerance (e.g., drought, salinity)	X	X	X
	Adaptation and productivity	X	X	X

TC/H, tissue culture foed by hybridization; MAS, Molecular Marker Assisted Selection; and Transgllow, Transgenics.

agriculture are becoming greater. Thus, biotechnology will continue to be gradually incorporated into routine breeding programs to improve accuracy, speed, and efficiency. Although the challenges are great, the prospects are even greater.

REFERENCES

Allard, R.W., 1971. Princípios do melhoramento genético das plantas. Traduzido por: Blumenscheub, A., Paterniani, E., Gurgel, J.T.A., Vencovski, R. Edgard Blücher, São Paulo Ltda, 381p.

Beddington, I., 2010. Food security: contributions from science to a new and greener revolution. Philosophical Transactions of the Royal Society 365, 61–71.

Belicuas, P.R., Guimarães, C.T., Paiva, L.V., Duarte, J.M., Maluf, W.R., Paiva, E., 2007. Androgenetic haploids and SSR markers as tools for the development of tropical maize hybrids. Euphytica 156, 95–102.

Bernardo, R., 2010. Breeding for quantitative traits, second ed. Stemma Press, Minneapolis. 400p.

Bernatzky, R., Tanksley, S.D., 1986. Toward a saturated linkage map in tomato based on isozymes and random cDNA sequences. Genetics 112, 887–898.

Borém, A., Caixeta, E.T., 2009. Marcadores moleculares, second ed. Editora Suprema, Visconde do Rio Branco. 532p.

Borém, A., Miranda, G.V., 2013. Melhoramento de plantas, sixth ed. Editora UFV, Viçosa. 523p.

Borlaug, N.E., 1968. Wheat breeding and its impact on world food supply. In: Public Lecture at the Third International Wheat Genetics Symposium. Australian Academy of Science, pp. 5–9.

Borlaug, N.E., 1969. A Green Revolution yields a golden harvest. Columbia Journal of World Business 4, 9–19.

Clay, J., 2011. Freeze the footprint of food. Nature 475, 287–289.

Cobb, J.N., DeClerck, G., Greenberg, A., Clark, R., McCouch, S., 2013. Next-generation phenotyping: requirements and strategies for enhancing our understanding of genotype–phenotype relationships and its relevance to crop improvement. Theoretical Applied Genetics 126, 867–887.

Duvick, D.N., 1986. Plant breeding: past achievements and expectations for the future. Economic Botany 40, 289–297.

Frankel, O.H., 1958. The dynamics of plant breeding. Journal of Australian Institute of Agriculture Science 24, 112.

Helentjaris, T., Writght, S., Weber, D., 1986. Construction of a genetic linkage map in maize using restriction fragment polymorphisms. Maize Genetics Cooperation Newsletter 60, 118–120.

Kubicek, Q.B., 1997. Panorama da biotecnologia nos EUA. Biotecnologia, Ciência e Desenvolvimento 1, 38–41.

Lande, R., Thompson, R., 1990. Efficiency of marker-assisted selection in the improvement of quantitative traits. Genetics 124, 743–756.

Lee, M., 1995. DNA markers and plant breeding programs. Advances in Agronomy 55, 265–344.

Milach, S.C.K., Cruz, R.P., 1997. Piramidação de genes de resistência às ferrugens em cereais. Ciência Rural 27 (4), 685–689.

Murovec, J., Bohanec, B., 2012. Haploids and doubled haploids in plant breeding. In: Abdurakhmonov, I. (Ed.) Plant Breeding. InTech, Available from: <http://www.intechopen.com/books/plant-breeding/haploids-and-doubled-haploids-in-plant-breeding>.

Rafalski, J.A., Tingey, S.V., Williams, J.G.K., 1991. RAPD markers, a new technology for genetic mapping and plant breeding. AgBiotech News and Information 3, 645–648.

Rommens, C.M., Kishore, G.M., 2000. Exploiting the full potential of disease resistance genes for agricultural use. Current Opinion in Biotechnology 11, 120–123.

Scott, K.D., Eggler, P., Seaton, G., Rossetto, M., Ablett, E.M., Lee, L.S., et al., 2000. Analysis of SSRs derived from grape ESTs. Theoretical and Applied Genetics 100, 723–726.

Sharma, H.C., Crouch, J.H., Sharma, K.K., Seetharama, N., Hash, C.T., 2002. Applications of biotechnology for crop improvement: prospects and constraints. Plant Science 163, 381–395.

Stoskopf, N.C., Tomes, D.C., Christie, B.R., 1993. Plant Breeding: Theory and Practice. Westview Press, Boulder. 300p.

Tanksley, S.D., McCouch, S.R., 1997. Seed banks and molecular maps: unlocking genetic potential from the wild. Science 277, 5329–5335.

Tanksley, S.D., Ganal, M.W., Martin, G.B., 1995. Chromosome landing: a paradigm for map-based gene cloning in plants with large genomes. Trends in Genetics 11, 63–68.

Vavilov, N.I., 1935. The origin, variation, immunity and breeding of cultivated plants. Ronald Press, New York. Tranduzido por K. Staer Chester. 1851.

Wenzel, G., 2006. Molecular plant breeding: achievements in green biotechnology and future perspectives. Applied Microbiology Biotechnology 70, 642–650.

Williams, W., 1964. Genetic Principles and Plant Breeding. Blackwell Press, Oxford.

Wolf, E.C., 1986. Beyond the Green Revolution. Worldwatch paper 73.

Chapter 2

Molecular Markers

Eveline Teixeira Caixeta,[a] Luis Felipe Ventorim Ferrão,[b]
Eunize Maciel-Zambolim,[b] and Laércio Zambolim[b]
[a]Embrapa Coffee, Brasilia, Brazil, [b]Federal University of Viçosa, Viçosa, Brazil

INTRODUCTION

Genetic breeding programs of plants have had a significant impact on agricultural production. Despite the success of these programs, additional gains in breeding efficiency can be obtained through the application of molecular technologies. Prominent among the available molecular technologies is the analysis of DNA polymorphism, which has become an area of active research in major agronomic species and model plants. Molecular markers defined in this context have proven to be useful in plant breeding and in increasing our understanding of crop domestication, plant evolution, and the genetic mechanisms involved in agronomical traits.

Several types of molecular markers have been described in the literature and are available to researchers. One of the greatest challenges in using such markers is choosing the types of molecular markers that best suit the purpose of the specific project. To succeed in the choice of marker, the first step is to be well acquainted with the types of marker available for the intended crop. To assist in this task, this chapter addresses the currently most frequently used markers in plant species. To facilitate the reader's understanding, the markers will be divided into three groups according to the method of analysis: (a) PCR-based markers; (b) hybridization-based markers; and (c) sequence-based markers.

PCR-BASED MARKERS

PCR (polymerase chain reaction) is a molecular biological technique that consists of the amplification of DNA fragments *in vitro* by the action of a replication enzyme. The basic procedure of PCR involves the following steps: (a) denaturation of double-stranded DNA by exposure to high temperature (typically 92 to 95 °C), separating it into single strands that will serve as templates for the amplification; (b) hybridization of a small fragment of single-stranded DNA, called primer, to the template (the hybridization occurs due to the presence of a complementary base sequence); (c) synthesis of a second strand complementary

Biotechnology Applied to Plant Breeding. http://dx.doi.org/10.1016/B978-0-12-418672-9.00002-7

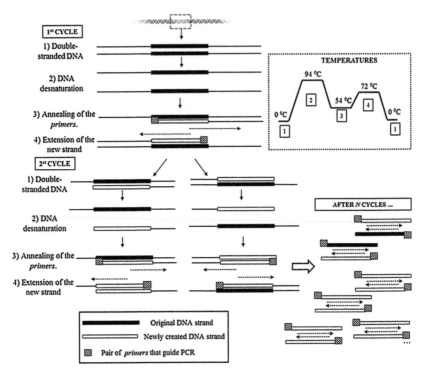

FIGURE 2.1   Diagram of DNA fragment amplification by PCR.

to the DNA template by DNA polymerase enzyme; (d) denaturation of the formed double-stranded DNA fragment; and (e) initiation of a new cycle. The cycle is repeated until millions of new DNA molecules are formed (Figure 2.1).

Numerous PCR-based markers that are suitable for use in various plant species are available. We will describe below the markers that are currently most used.

Microsatellite Markers

Genetic Basis of Microsatellite Markers

Microsatellites, also called SSR (simple sequence repeats) and STR (short tandem repeats), are repeated sequences that occur in *tandem* and that vary from 1 to 6 base pairs (bp) in length. Microsatellites are classified based on their size, type of repeat unit, and location in the genome (Table 2.1). Abundant and well distributed in eukaryotes, SSR are distinguished by wide variations in the repeat number resulting from dynamic and complex mutagenic events such as unequal *crossing over*, retrotransposition, and DNA polymerase *slippage* (Ellegren, 2004; Caixeta et al., 2009; Kalia et al., 2010).

The use of microsatellites as molecular markers is possible due to the great polymorphism of repetitive regions resulting from mutagenic events coupled

TABLE 2.1 Classification of Microsatellites

Size	Repeat unit	Location
Mononucleotide (A)$_n$	Simple perfect (CA)$_n$	Nuclear (nuSSRs)
Dinucleotide (CA)$_n$	Simple imperfect (AAC)$_n$ACT(AAC)$_n$	Chloroplast (cpSSRs)
Trinucleotide (CGT)$_n$	Compound perfect (CA)$_n$(GA)$_n$	Mitochondria (mtSSRs)
Tetranucleotide (CAGA)$_n$	Compound imperfect (CCA)$_n$TT(CGA)$_n$	
Pentanucleotide (AAATT)$_N$		
Hexanucleotide (CTTTAA)$_n$		
(n = number of repetitions)		

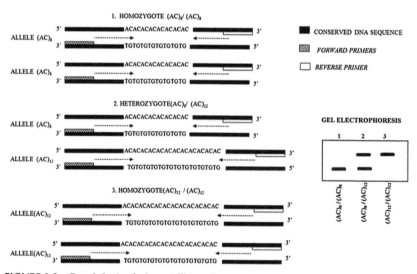

FIGURE 2.2 Genetic basis of microsatellite markers.

with the fact that the DNA sequences flanking these repetitions are normally preserved between individuals of the same species. This feature allows the design of specific primers (20 to 30 bp in length) for the sequences adjacent to the polymorphic microsatellite, so that it is possible to amplify this region by means of a PCR reaction. As illustrated in Figure 2.2, amplification products feature polymorphisms that differ in fragment size, thus allowing for the differentiation of genotypes. In this way, each microsatellite constitutes a highly variable, multi-allelic genetic locus that has great informative content because it permits differentiation between homozygous and heterozygous organisms (codominant expression) (Ferreira and Grattapaglia, 1998).

These characteristics of microsatellites, together with the high reproducibility and simplicity of PCR, make microsatellites the most widely used markers not only in plant breeding but also in ecology, conservation biology, and phylogenetic studies (Selkoe and Toonen, 2006). In plant breeding, in particular, the use of microsatellite markers stands out in genetic mapping studies, in the identification of QTLs (quantitative trait loci), in diversity studies in germplasm banks, in marker-assisted selection (MAS), and, recently, in studies of wide genomic selection. The major limitation of this class of markers lies in the need to isolate and develop primers specific to each species, which can be a toilsome, lengthy, and costly process. However, this task has been facilitated by the sequencing of DNA from different species, which resulted in the availability of thousands of sequences in public and private DNA databases. Knowledge of these sequences and their use in concert with new bioinformatics techniques have allowed the development of primers for SSR without the need for the complex initial steps of the traditional strategy, making the process far less laborious.

Detection and Analysis of Polymorphisms in Microsatellites

To detect polymorphisms based on microsatellite markers, the use of methodologies that separate amplified fragments by molecular size is required. In general, gel electrophoresis and capillary electrophoresis techniques by automated systems are used. In both cases, each microsatellite locus is analyzed individually; the amplified fragments represent different alleles of the same locus.

Resolution of DNA fragments of various sizes by electrophoresis is based on the movement of molecules through a solid matrix. The matrix acts as a filter, separating the molecules according to size and electrical load. Electrophoresis in agarose gels is a standard method for separating, identifying, and purifying DNA fragments. Simple, practical, and inexpensive, this technique allows the direct visualization of DNA fragments after the use of specific stains (usually ethidium bromide). In studies of microsatellite, the use of agarose is restricted due to the low resolution of fragment separation. To minimize this limitation, high concentrations (3–5%) of special agarose have been used (Figure 2.3); the trade names of some of these products are MetaPhor (FMC-Bioproducts), NuSieve, and SFR (AMRESCO).

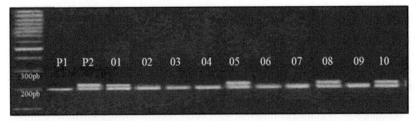

FIGURE 2.3 Polymorphism detection in MetaPhor agarose gel. Genotypic mapping of a population of cotton resulting from backcrossings (P1 × P2). (*Adapted from Asif et al., 2008*)

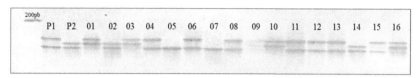

FIGURE 2.4 **Polymorphism detection in a 6% polyacrylamide denaturing gel stained with silver nitrate.** Genotypic mapping of a *Coffea canephora* population resulting from the crossing of two heterozygous parents (P1 × P2).

Polyacrylamide gel electrophoresis is the most commonly used approach for the identification of microsatellites because it combines low cost with high resolution, permitting the separation of DNA fragments that differ in length by only 1 bp. The amplified fragments can be detected by staining with ethidium bromide or silver nitrate, by fluorescence, or by incorporation of ^{32}P or ^{33}P, allowing accurate visualization of the bands. Denaturing agents (usually urea) are sometimes used to increase the discriminatory power of these gels. This strategy improves the resolution of the fragments and is advisable when attempting to distinguish alleles of proximate sizes (Figure 2.4).

Another strategy used to detect PCR products is separation of the products by capillary electrophoresis. Comparative studies of electrophoretic techniques show that the resolution and accuracy of this methodology for the differentiation of fragments of similar sizes are superior to those of gel electrophoresis. In this case, polymorphism identification is based not on differentiation of the band size but on the intensity of the fluorescent peaks (Figure 2.5). To achieve this, it is necessary to perform PCR with previously marked primers that can be identified during electrophoresis. A standard fluorescence marker is added during the electrophoresis run to serve as a parameter for calculation of the size of the amplified fragment. Thus, at the end of each electrophoretic run, polymorphism among individuals is represented by the presence of multiple peaks, and it is possible to quantify the difference in the number of base pairs present in the amplified alleles.

Because capillary electrophoresis is very sensitive, successful use of the technique requires optimization of each step in the process. Among the standard procedures for optimization, we emphasize the balanced use of DNA and primers, the use of an appropriate annealing temperature, fluorescence testing of the markers and prior analysis of the amplification profile of each microsatellite by means of high-resolution polyacrylamide gels. Such care can avoid problems such as *stutters* (ghost peaks), double peaks (*split peaks*), low peaks in heterozygotes, null alleles, and nonspecific amplification. These artifacts, which are illustrated in Figure 2.6, make it difficult to read the electropherograms and create doubts about the validity of the evaluation. *Stutter* corresponds to the amplification of PCR products that differ from the target DNA by one or a few repetitions. *Stutter* can be reduced by lowering the denaturation temperature or by using new-generation polymerase enzymes. Double peaks often result from the addition of a non complementary nucleotide (usually adenine) to the

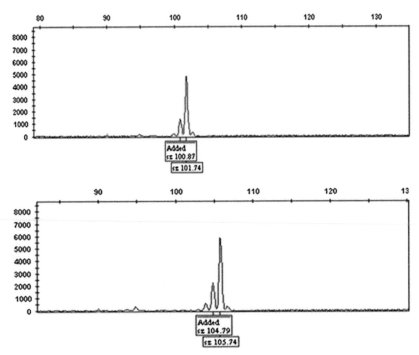

FIGURE 2.5 Polymorphism detection by capillary electrophoresis using fluorescent labeled primers. Electropherogram obtained from automatic sequencer.

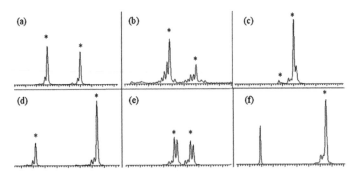

FIGURE 2.6 Microsatellite marker profiles generated by capillary electrophoresis. (a) Correct profile; (b) profile with *stutters*, small peaks that appear before the true peak; (c) null allele; (d) low peak of heterozygote; (e) double peaks; (f) artificial peak resulting from nonspecific primer pairing. The peaks corresponding to the correct alleles are marked with asterisks (*). (*Figure adapted from Guichoux et al., 2011*)

amplified fragment by *Taq* polymerase. If the addition does not occur in all fragments, there may be two peaks: one corresponding to the fragment with the adenine and the other to the fragment without the adenine. This problem can be minimized by increasing the final extension time of the reaction.

Alternative solutions involve adjusting the concentrations of the PCR reagents by increasing the concentration of polymerase, reducing the amount of target DNA, or reducing the concentration of primers. Low peaks in heterozygotes and the appearance of peaks corresponding to null alleles are caused by mutations in the regions flanking the microsatellites at the primer annealing site. In this case, if the microsatellite is important, the primers should be eliminated or redesigned to avoid the polymorphic region.

Microsatellite markers are highly informative and locus specific; however, only a single locus can be analyzed per assay. To increase the speed and efficiency of genotyping using these markers, multiplex reactions are an option. Multiplex PCR allows multiple markers (primers) to be simultaneously amplified in the same reaction, increasing the amount of information generated per assay. However, designing a robust multiplex PCR test is not simple due to the possibility of undesirable interactions between the different primers that are used. These interactions may result in primer pairing, which prevents the amplification of the DNA template. Therefore, multiplex PCR tests should be designed with caution, selecting primers that are as similar as possible in size and in amplification conditions, and that have been tested for the possibility of interaction (Guichoux et al., 2011).

Another option for the simultaneous analysis of more than one locus is the use of a pseudo-multiplex strategy. In this case, the PCR reactions are run separately and then mixed together for electrophoresis. Because the reactions occur independently, the difficulties mentioned in the case of optimization of multiplex reactions are minimized.

After obtaining the data set by gel or capillary electrophoresis, the next step is the determination of the genotype of each individual. For this, either of two methods may be employed. The first, which is used most often following gel electrophoresis, consists of converting the sizes of the alleles into numerical units or labels (Figure 2.7a). The other method, which is more often used after capillary electrophoresis, designates each peak by the actual size of the allele;

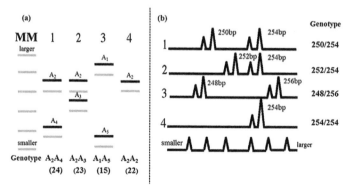

FIGURE 2.7 (a) Allele identification for a microsatellite locus of a diploid organism from bands in a polyacrylamide or agarose gel. (b) Genotype identification based on the electropherograms produced by automated equipment. MM: marker of known molecular size.

thus, the denomination of the allele corresponds to the number of existing base pairs (Figure 2.7b). In the example shown, the analysis of diploid and heterozygote genotypes with alleles that are 248 bp and 256 bp in length can be made by discrimination of genotypes using these values or by converting them into numerical units (1 5) or labels (A_1A_5).

Identification and determination of the sizes of alleles from data obtained using automated equipment (Figure 2.7b) can be made with the aid of commercial software designed to facilitate the analysis of the electropherograms. However, the size of each allele revealed by gel electrophoresis can only be determined based on a stochastic method, i.e., by comparing the obtained fragment with a marker of known molecular weight. Thus, labeling or numbering of alleles is a simpler and possibly more viable alternative (Figure 2.7a).

Frequent and well distributed throughout the genome, microsatellites are distinguished by being multi-allelic, codominant, and highly reproducible markers (Ferrão et al., 2013). These attributes make microsatellites one of the most commonly used markers in plant breeding; they are especially useful in diversity, mapping, and assisted selection studies (Varshney et al., 2005). The recent development of new technologies for sequencing, combined with the ease of use of multiplex reactions, promises to significantly increase the use of these markers by researchers.

ISSR Markers

Another PCR-based marker that has been widely used is the ISSR (inter simple sequence repeat). These molecular markers are based on the amplification of DNA fragment located between two identical microsatellite regions (Figure 2.8). This technique uses the repeated DNA sequence (SSR) as a single primer in PCR reactions, thus allowing the simultaneous amplification of different regions of the genome. It is a simple and rapid technique that combines comprehensive coverage of the genome and reproducibility with the ability to analyze a greater number of loci per assay. In addition, ISSR markers use universal primers, i.e., random primers whose sequences are independent of the species under study, making this technique appropriate for studies that involve the analysis of several

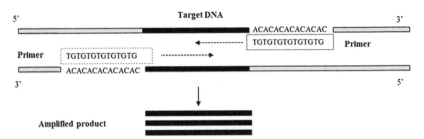

FIGURE 2.8   The genetic basis of ISSR markers. The PCR reaction uses a single primer of microsatellite sequence (TG)$_7$.

species. The vast majority of ISSRs segregate as dominant markers, which constitutes the greatest limitation to their use. In plant breeding, ISSR markers are commonly used in diversity, mapping, and assisted selection studies (Reddy et al., 2002).

AFLP Markers

Genetic Basis of AFLP Markers

Described by Vos et al. (1995), AFLP (amplified fragment length polymorphism) markers stand out because they permit the simultaneous analysis of different DNA regions that are randomly distributed in the genome. The technique for using these markers consists essentially of four steps and is based on selective amplification of a subset of genomic fragments generated after digestion of genomic DNA with restriction enzymes (Figure 2.9). This technique combines universal applicability with high discriminatory power and reproducibility, making it an important tool in genetic studies of crops that have low rates of DNA polymorphism.

The first and second steps of the AFLP marker technique consist, respectively, of the digestion of genomic DNA with restriction enzymes and the ligation of sequence adapters to these fragments (Figure 2.9 A and B). At this stage, two restriction enzymes are usually used, one rare cutter (e.g., *EcoRI*) and one frequent cutter (e.g., *MseI*). Double-stranded DNA adapters are then ligated to the ends of the digested fragments. These adapters are 20 to 30 bp in length and serve as sites for primer annealing and amplification via PCR.

Subsequent steps in the AFLP technique include the selection of a subpopulation of fragments that can be conveniently analyzed on a gel. In this third step, fragments to be amplified by means of a PCR reaction with two different primers are selected. In addition to containing the complementary sequence specific to the adapter, these primers have, in general, an additional nucleotide at their 3′ ends. In this way, only fragments that have the selective nucleotide are amplified. Because this is the first selection made in the amplification, it is called "pre-selective amplification" or "pre-amplification." In the next step, "selective amplification," further amplification by PCR is performed using primers consisting of the adapter sequences with three additional random nucleotides at their 3′ ends. The first of these three nucleotides is the same as that used in the pre-amplification. Thus, a second selection of amplified fragments occurs (Figure 2.9C).

The last step consists of analyzing the subset of amplified fragments by gel or capillary electrophoresis (Figure 2.9D). he advantage of this technique is the large number of generated fragments, which can be analyzed in a single assay. Thus, these markers are frequently chosen for genetic diversity and mapping studies of plants that have low rates of polymorphism. However, AFLP markers present difficulties in the identification of allelic variants at a specific locus because the technique lacks sufficient sensitivity to discriminate heterozygous

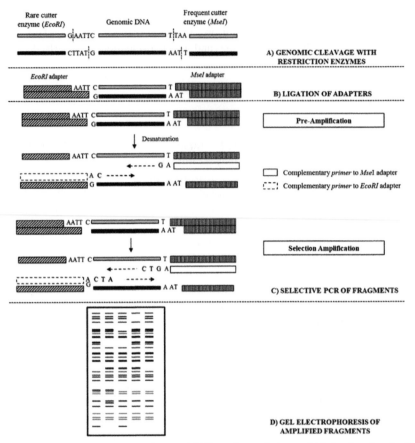

FIGURE 2.9 **Diagram representing the AFLP marker technique, which involves four basic steps**. (A) DNA digestion; (B) ligation of specific adapters; (C) fragment selection via PCR; and (D) gel electrophoresis for visualizing polymorphisms.

genotypes. Furthermore, AFLP involves a greater number of steps, making the technique more laborious.

Polymorphism Detection and Analysis in AFLP

The detection of AFLP markers involves visualization of the size differences between the generated fragments. As in SSR, AFLP products can be analyzed by gel electrophoresis or by an automated capillary electrophoresis instrument.

When AFLP markers are used, agarose gel electrophoresis is recommended only in the early stages to confirm that the digestion and pre-amplification steps occurred correctly. For the final analysis of the generated fragments, the best resolution of the bands is obtained with the use of polyacrylamide denaturing gels at concentrations of 4 to 6% acrylamide; these gels resolve fragments that differ in a single nucleotide (Figure 2.10A). Polymorphism can be visualized by

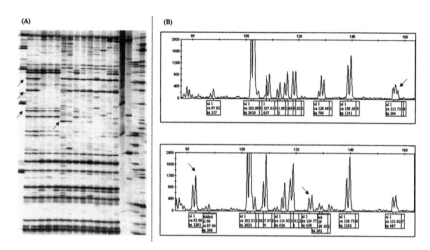

FIGURE 2.10 (A) Polymorphism detection using AFLP markers, polyacrylamide denaturing gel electrophoresis, and colorimetry with silver nitrate; (B) capillary electrophoresis with labeled primers. The arrows indicate polymorphisms.

staining the gel with silver nitrate or by autoradiography of radioactive labeling primers. Comparative studies indicate that the sensitivity and the resolution of these two methods are similar; however, staining with silver offers a significant advantage in that it allows fragments of interest to be isolated directly from the gels (Cho et al., 1996; Chalhoub et al., 1997). In this case, the bands of interest can be removed from the gel and immersed in rehydration solutions for subsequent characterization and more detailed studies (Blears et al., 1998). This procedure allows the conversion of specific bands into highly reliable markers, thus enabling the development of another category of molecular markers called SCAR (sequence-characterized amplified region).

The use of automated capillary electrophoresis systems permits the analysis of a large number of loci. The resulting electropherograms are analyzed using specific software that provides the size of each allele in base pairs (Figure 2.10B). As in microsatellite, reactions are conducted using primers labeled with specific fluorophores, thus permitting the detection of polymorphisms by an automatic sequencer. Standard fluorescence markers (ROX or LIZ) should be used for the calculation of the sizes of alleles.

In the evaluation, it is assumed that each peak or band corresponds to an allele of a different locus. Because AFLP markers are typically dominant markers, heterozygous genotypes cannot be discriminated; therefore, evaluation is based on the presence (1) and absence (0) of bands/peaks, resulting in binary data (Savelkoul et al., 1999). The large number of bands or peaks observed in evaluations of AFLPs increases the "multiplex index" of these markers, i.e., the number of markers that can be analyzed simultaneously in a single electrophoretic run. However, the appearance of artifacts in this type of analysis is common; for this reason, it is advisable to consider only clear and easily distinguishable bands and peaks. Fragments of unclear or conflicting identity should be disregarded.

SCAR Markers

A SCAR marker can be defined as a fragment of genomic DNA that was identified by PCR amplification using a pair of specific primers (Paran and Michelmore, 1993). SCARs represent a category of molecular markers derived from other markers, such as RAPD and AFLP. For SCAR development, genomic DNAs are analyzed with RAPD or AFLP markers, and bands of interest, such as those linked to specific agronomic traits, are identified and isolated directly from the gel. These fragments are cloned and sequenced, thus permitting the design of specific primers flanking the region of interest. Use of these primers in PCR reactions offers the great advantage of converting dominant markers into codominant ones. SCAR markers are locus specific and offer higher resolution than other classes of markers.

HYBRIDIZATION-BASED MARKERS

The first hybridization-based marker developed was the RFLP (restriction fragment length polymorphism); this marker was first used in 1975. RFLP is based on the digestion of an individual's DNA with one or more restriction enzymes, separation of the DNA fragments by agarose gel electrophoresis, transfer of the fragments to a solid support (nylon or nitrocellulose membrane), and detection of the fragments by hybridization with a probe corresponding to a DNA with a known sequence. Although it offers very robust markers for genetic analyses and use in breeding programs, the limitations and difficulties of the RFLP technique have hampered its routine use, and the technique is not widely used currently. The primary limitations of RFLP are its high cost, the difficulty involved in the implementation of many steps, and the requirement for considerable investment in staff, equipment, chemicals, and special care when radioactive probes are used. These aspects of the techniques are also obstacles to its automation, making it difficult to generate a large volume of data.

In 2001, a technique called DArT (diversity array technology), which is based on hybridization by microarray, was proposed by Jaccoud et al. (2001). DArT was designed to provide a robust technique for large-scale analysis at lower cost, using the minimum DNA possible to provide extensive coverage of the genome, and without the knowledge of the species' DNA sequence.

DArT Markers

DArT markers allow the simultaneous genotyping of several hundreds of polymorphisms in large regions of the genome with high reproducibility. Generally, approximately 50–100 ng of genomic DNA is used to simultaneously genotype approximately 5000–8000 loci in a single assay. Thus, DArT can be used to characterize several hundreds to millions of polymorphisms relatively quickly and at an accessible cost per datum (*data point*) compared to the use of other molecular markers. DArT markers are biallelic and behave in a dominant way in the majority of cases.

DArT markers are derived through the construction of a DNA array containing a DNA *pool* of different genotypes (called a "diversity panel"), followed by hybridization with a probe derived from DNA extracted from individual genotypes (Figure 2.11). To create the "diversity panel" (Figure 2.11A), DNAs from different individuals who are representative of the species under study are extracted, pooled, and digested with restriction enzymes. Different restriction enzymes can be used; usually two enzymes, a rare cutter and a frequent cutter, are selected for each test. The choice of enzyme combination is carried out individually for each species to optimize the reduction in genome complexity. The resulting fragments are linked to specific adapters. The genome complexity is reduced by PCR using selective primers containing sequences complementary to the adapters and selective extensions involving the addition of one, two, or three bases at the 3′ end. Reduction in the complexity of the DNA sample results in the formation of a "genomic representation" of the species. The DNA fragments of the "genomic representation" are cloned in specific vectors and transformed into bacteria, resulting in their representation in the form of clones in a genomics library. After purification, the fragments are arrayed and attached to a solid support (glass slide), forming a DNA microarray that corresponds to the "diversity panel." Two or three replicas of each clone are deposited in the microarray.

The principle of DArT is that the "genomic representation" contains two types of DNA fragments, common fragments found in any "representation" prepared from a DNA sample of an individual belonging to the species as well as polymorphic fragments (the DArT markers) that are found only in some genotypes.

The diversity panel (microarray) is hybridized with labeled probes prepared from each individual's DNA (Figure 2.11B). The individuals from whom the probes are derived should preferably have been included in the "genomic representation." To obtain the probes, each individual's DNA is subjected to the same steps of complexity reduction used to obtain the "genomic representation" and labeled with a fluorescent dye. Each individual probe, also known as the "target," is mixed with *polylinker* fragments of the vector used in the construction of the library. These vector fragments are labeled with a different fluorescent dye and will serve as a DNA reference for quantifying the DNA deposited in the array. The diversity panel is hybridized with both the target and the reference probes, and the intensity of each fluorescent dye is measured by the laser scanner using appropriate wavelengths. The obtained data are analyzed using specific software such as DArTsoft (http://www.diversityarrays. com/software.html) and DArTdb Laboratory Information Management System

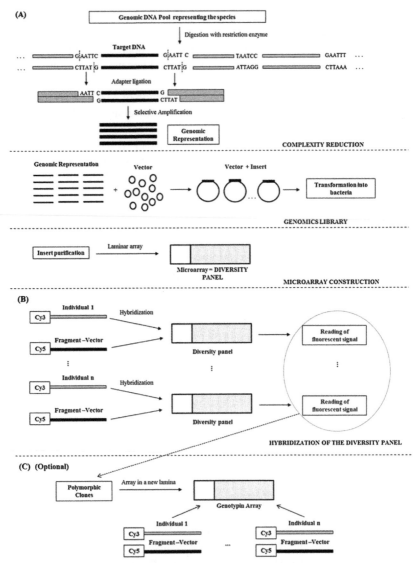

FIGURE 2.11 DArT marker technique. (A) The genomic DNA pool is cleaved by restriction enzymes. The resulting fragments are linked to adapters and amplified by PCR using selective primers. Fragments of the genomic representation are cloned, purified, and arrayed on a solid support, forming the diversity panel. (B) The DNAs of the individuals to be genotyped are subjected to complexity reduction using the same methodology used to obtain the genomic representation. The fragments are labeled with fluorescent dyes and used as probes for hybridization in the diversity panel. A fragment of the vector labeled with another fluorescent dye serves as a reference in the hybridization. (C) For some species, another panel, the genotyping array, is built before making the genotyping of the population. This allows the DNA of some individuals belonging to the DNA pool to be used as a probe and hybridized in the diversity panel. Clones that present polymorphism are arrayed in a new solid support, and this array is used to analyze all the individuals of the population.

(http://www.diversityarrays.com/software.html%2523dartdb). Polymorphic clones (DArT markers) present variable hybridization signals for different individuals.

In another slide, polymorphic clones are arrayed, forming the "genotyping array" that will be used in routine genotyping (Figure 2.11C). In routine genotyping, representations of each individual to be analyzed are labeled and hybridized in the "genotyping array," and data from all of them are compiled, compared, and analyzed using the appropriate software. Alternatively, individual representations are hybridized directly on the diversity panel; in this case, the genotyping array is not built.

The technique described above is a general methodology for DArT identification and genotyping. However, some modifications to the technique, such as using different strategies to reduce the genome complexity and including an enrichment step for polymorphic clones, as in the SSH (suppression subtractive hybridization) technique, can be made. For some species, the construction of the genotyping array is performed using polymorphic clones from different diversity panels, with each panel being obtained from different groups of genotypes. Another variation is the use of three probes per hybridization, which enables the analysis of the reference probe and two targets in a single hybridization test.

DArT analysis does not require prior sequence information on the species. In addition to providing high yield, high reproducibility, and a relatively quick methodology, this makes the technique applicable to any species. However, because it is based on microarray hybridization, it involves several steps, including preparation of the genomic representation of the target species, cloning, hybridization, and data analysis. Therefore, the establishment of the DArT system demands a great investment in physical structure and personnel. Once established, the technique has a relatively low cost per data point compared with other markers. The cost per data point is 10-fold lower than for SSR.

MARKERS BASED ON SEQUENCING

DNA sequencing consists of determining the order of the bases of the nucleotides adenine (A), guanine (G), cytosine (C), and thymine (T) present in a DNA molecule. There are several techniques available for sequencing. The technique that is considered the conventional technology for DNA sequencing is called the Sanger method. This technique is based on PCR in which an interruption of the sequence occurs through the random incorporation of a modified nucleotide that is labeled with a fluorescent dye. DNA synthesis is initiated by DNA polymerase following the annealing of a primer to the DNA template. The polymerase synthesizes a new strand from the sequence template, incorporating the nucleotides. Two types of nucleotides are added to the reaction: deoxynucleotides (dATP, dGTP, dCTP, and dTTP) and dideoxynucleotides (ddATP, ddCTP, ddGTP, and ddTTP). The dideoxynucleotides lack hydroxyl groups on their 3′ carbon atoms and are typically marked with fluorescent dyes. DNA polymerization continues with the inclusion of the "normal" nucleotides (deoxynucleotides) until, by chance, the DNA polymerase inserts a dideoxynucleotide, which interrupts the polymerization. As a result of this process, fragments of different sizes are obtained. The reaction is then placed in an automatic sequencer, which identifies the fragments by *laser* spectroscopy of the fluorescent dideoxynucleotides. The Sanger method is an automated, robust, highly accurate (>98%) technology that can sequence fragments up to 900 bp in length.

Despite the prevailing use of the Sanger method for DNA sequencing, new methods called "next generation sequencing" or "second generation sequencing" are emerging. These new methods leverage the development and use of molecular markers based on DNA sequencing and provide DNA sequencing in platforms capable of generating information on millions of base pairs in a single assay. These platforms sometimes have greater power to generate information than Sanger sequencing, with great savings in time and cost per base. Their greater efficiency arises from the use of *in vitro* cloning and solid support systems for the sequencing units. Some examples of next generation sequencing platforms include 454 FLX by Roche, Solexa by Illumina, and SOLiD System by Applied Biosystems (Liu et al., 2012). These techniques have made it possible to perform genetic studies of populations based on complete genomes instead of on short sequences of a single gene.

The first next-generation sequencing platform developed was the 454 FLX; this type of platform is also sometimes referred to as pyrosequencing. The DNA fragments sequenced in this platform are usually 250 bp or less in length, much shorter than the sequences produced by the Sanger system. With use of the *Titanium* series pyrosequencing methods, it is possible to obtain readings up to 400 bp; each assay permits the sequencing of approximately 0.5 Gb of genomic DNA. Another frequently used platform is Solexa, which can sequence fragments that are 25 to 35 bp in length and allows the sequencing of 3 Gb of genome per assay. In the SOLiD platform, fragments that are approximately 35 to 50 bp in length are sequenced, and 25 Gb can be sequenced in an assay. The latter two platforms are available at lower cost than the 454 FLX, but the process of genome assembly using the smaller fragments generated by Solexa and SOLiD is more laborious, especially in *de novo* sequencing projects that involve the sequencing of unknown genomes.

With the advent of next-generation sequencing technologies, markers such as SNP (single-nucleotide polymorphism) and genotyping by sequencing have been developed and used in different plant species. These markers provide an approach to a wide analysis of the genome.

SNP Markers

SNP detection is based on the variation of DNA sequences in a single nucleotide, i.e., on changes of a single base in a specific position of the genome (Figure 2.12). SNPs can be classified according to the nucleotide variation; they may be transitions, exchanges of a purine for another purine (A/G) or a pyrimidine for another pyrimidine (C/T), or transversions, exchanges of a

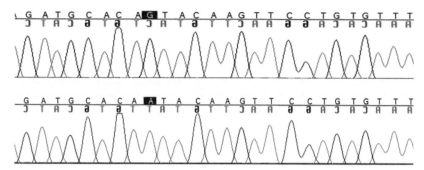

FIGURE 2.12   Identification of SNPs in different individuals by alignment of DNA sequences.
The SNP in the example corresponds to the replacement of G with A.

purine for a pyrimidine or a pyrimidine for a purine (A/C, A/T, G/C, G/T). Transitions are the most frequent variations.

Single-nucleotide variations may occur in coding regions or in regions with regulatory functions, but most of them are found in non coding regions of the genome. Variations in coding regions are classified as synonymous or non synonymous. Non synonymous mutations result in a change of the amino acid encoded in the protein sequence; this change may be conservative or non conservative based on the characteristics of the amino acid specified by the exchange. Polymorphisms that produce non conservative changes may lead to structural changes in the encoded protein and consequently to changes in its function. Mutations of the synonymous type are those in which the presence of the polymorphism does not result in an alteration of the encoded amino acid.

SNPs and small insertions and deletions (*indels*) represent the most frequent natural genetic variation in most organisms and have been used as next-generation molecular markers in various applications. Markers based on SNPs are generally biallelic, i.e., they have two alleles per locus; they are therefore less informative than SSRs, which are usually multi-allelic (having multiple alleles per locus). This disadvantage is compensated for by the abundance and capacity of the SNPs to be used in large-scale genotyping (*ultra-high-throughput genotyping*).

As for most markers, it is necessary to identify and validate SNPs for the species before beginning the genotyping of populations. The most direct way to identify polymorphisms in DNA is to directly sequence PCR products from a number of divergent individuals. The primers for the PCR are designed based on known DNA sequences, usually the sequences of genes of interest. In general, the primers are designed in such a way that their 5′ ends do not anneal to the DNA template but instead link to a sequencing primer. After the sequencing of DNA fragments, the sequences are compared, and SNPs are identified. It is not necessary to limit the search for SNPs to sequences present in known genes. Especially for species with less polymorphism, search for SNPs can be more productive in regions that are not part of a gene.

Another way to identify SNPs is by an *in silico* search in which EST (Expressed Sequence Tag) databases are analyzed and compared. Because ESTs are normally derived from a number of distinct genotypes, many polymorphisms are expected in ESTs that represent the same gene. In this way, it is possible to identify a large number of SNPs through computer analysis. The greatest difficulty with this strategy is distinguishing between true polymorphisms and sequencing errors. Some software is being developed to assist in this task. After SNPs are identified *in silico*, it is necessary to experimentally validate them through analysis of a set of different genotypes.

Once identified and validated, SNPs must be genotyped in a significant number of individuals to be useful in genetic and plant breeding studies. A variety of SNP genotyping platforms are available (Gupta et al., 2008), and these differ in efficiency and cost. Unlike markers based on PCR, the use of SNP markers does not require separation of DNA fragments by size, so it can be performed on plates with 96 or 384 wells. Methods for higher-density assay have also been developed. The steps used in the various platforms include reactions that permit allele discrimination followed by reactions that allow allele detection. Allelic discrimination reactions are based on one of the following: (a) oligonucleotide hybridization; (b) primer extension; (c) oligonucleotide ligation; (d) cleavage with nuclease; and (e) direct sequencing.

In oligonucleotide hybridization, two allele-specific oligonucleotide probes are designed such that the polymorphic base is in a central position in the probe sequence. Under optimized conditions, only hybrids with perfect pairing between the probe and the target DNA are stable; hybrids with an unpaired base are unstable. In this strategy, the DNA is fixed in a membrane or other solid support, such as glass, plastic, or silicon, forming a DNA chip. The chip is hybridized to the probes, using a DNA chip for each probe. Only perfect pairing is detected; it indicates the presence of the SNP-allele in the studied DNA (Figure 2.13). Because it does not require enzyme activity, hybridization is the simplest method used in genotyping. The major limitation of this strategy is that hybridization is subject to errors because it requires well-designed probes and optimized protocols. The stability of the probe-DNA hybrid containing the

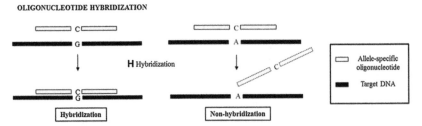

FIGURE 2.13 Allelic discrimination using oligonucleotide hybridization. The hybridization of a probe to different DNA samples is shown. Only the DNA-probe hybrid with perfect pairing is stable; the hybrid containing a non paired base is unstable. After washing, only the stable hybrid remains and is detected. In this example, individuals containing the G-allele are detected. The A-allele is detected analogously in a parallel reaction.

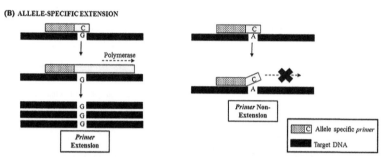

FIGURE 2.14 **Allelic discrimination of a SNP (G/A) by primer extension.** (A) Minisequencing: primer ligation to the DNA template in the presence of labeled ddCTP complementary to the G base, dNTPs of other bases and polymerase enzyme. In the reaction, the polymerase adds ddNTP or dNTP, introducing a base that is complementary to that in the target DNA. When the G-allele is present, labeled ddCTP is incorporated at the corresponding position, the polymerization reaction is completed, and a fragment of the size of the primer + one base (31 bp) is generated. For individuals containing the A-allele, another dNTP is incorporated, and the reaction continues until a G appears in the template and ddCTP incorporation stops the reaction. In this example, individuals who amplify fragments 31 bp in size feature the G-allele, and those who amplify fragments 35 bp in size feature the A-allele. (B) Allele-specific extension: DNA polymerase only extends the primer when it is perfectly complementary to the target DNA (C-primer).

corresponding SNP is determined not only by the stringency of reaction conditions but also by the sequence of nucleotides flanking the SNP and the secondary structure of the target sequence.

Primer extension is the most robust mechanism for allelic discrimination. This methodology is more flexible than oligonucleotide hybridization, and designing the primer and optimizing the reaction conditions are usually simple. The technique is based on the ability of DNA polymerase to incorporate nucleotides that are complementary to the sequence of the target DNA. There are a number of variations of the primer extension reaction, which are represented by two main types of reaction. The first is the minisequencing reaction, in which polymorphism is determined by the addition of a nucleic acid analog containing some modification, for example, a dideoxynucleotide triphosphate (ddNTP) (Figure 2.14A). In this case, the reaction includes a specific primer complementary to the region preceding the SNP intended for genotyping, fluorescently labeled dNTPs complementary to one of the SNP bases, and dNTPs of the other bases. The primer binds to the DNA template, and polymerase adds the ddNTP or dNTP to the 3′ end of the primer to introduce a base that is complementary

OLIGONUCLEOTIDE LIGATION

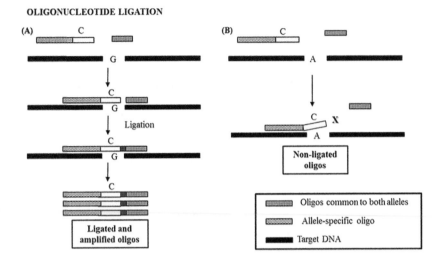

FIGURE 2.15   **Allelic discrimination by oligonucleotide ligation.** (A) Perfect pairing of both the allele-specific and common oligonucleotides results in the ligation of two DNA fragments and product detection by PCR. (B) When there is imperfect pairing of the allele-specific oligonucleotide with the DNA template, ligation and fragment formation do not occur.

to the base in the target DNA. For one of the SNP alleles, the first incorporated basis is the labeled ddNTP. After the incorporation, the polymerization reaction is complete, and a fragment is generated that is the size of the primer with a single added base. When DNA from individuals containing the other allele is used, dNTPs are incorporated and the reaction continues until ddNTP incorporation interrupts the reaction. Thus, individuals who present one of the alleles will form fragments with sizes corresponding to that of the primer plus one base, while fragments derived from the other allele will be larger (Figure 2.14A). The other main type of reaction is allele-specific extension, in which two primers, one containing a base complementary to one of the SNP alleles at its 3′ end and another containing a base complementary to the other allele, are required. Primer extension through DNA polymerase only occurs when the primer is perfectly complementary to the target DNA. The genotype of the sample is determined by detecting the primer responsible for the amplification of the product (Figure 2.14B).

Another reaction that can be used for allelic discrimination is oligonucleotide ligation. This method is based on the ability of DNA ligase to covalently link two oligonucleotides that are close to each other in the DNA template. In this case, three oligonucleotides are required, two of which are allele specific (one for each allele) and the other is common to both alleles. The common primer is annealed in a region of the DNA template in a region upstream of the SNP. Only the allele-specific primer that pairs perfectly with the DNA template will be ligated to the common primer by DNA ligase (Figure 2.15).

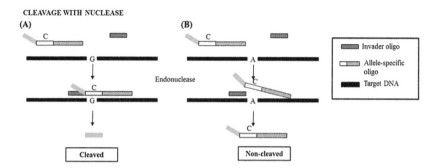

CLEAVAGE WITH NUCLEASE

FIGURE 2.16 Allelic discrimination by cleavage with nuclease. Pairing of allele-specific oligo-nucleotides and the invader with the target DNA forming an overlap at the 5″ end. (A) If the allele-specific oligonucleotide is complementary to the polymorphic base, the overlapping hybrids form a structure that is recognized and cleaved by endonuclease, releasing the 5′ fragment of the allele-specific oligonucleotide. (B) In the absence of the formation of the structure, there is no cleavage.

In allelic discrimination by cleavage with nuclease, two allele-specific oligonucleotides, one complementary to the SNP-allele and another, called the invader, complementary to the 5′-sequence upstream of the SNP allele, are required. When these oligonucleotides anneal with the target DNA, they overlap in the sequence corresponding to the invader. When the allele-specific oligonucleotide is complementary to the polymorphic base, the overlapping oligonucleotides form a three-dimensional structure that is recognized and cleaved by endonuclease, releasing the 5′ portion of the allele-specific oligonucleotide (Figure 2.16). The method of detection permits the identification of individuals for whom the reaction product corresponds either to the cleaved 5′ end of the oligonucleotide or to the uncleaved oligonucleotide.

After the discrimination of SNP alleles, precise detection is necessary. A variety of detection methods are available for analysis of the products generated by the allelic discrimination methods described here; these methods include, among others, fluorescence, FRET (Fluorescence Resonance Energy Transfer), fluorescence polarization, chemiluminescence, mass spectrometry, and chromatography. Analysis of the reaction products of allelic discrimination can be performed using one or more detection methods, and the same detection method can be used to analyze products obtained from different reactions (Figure 2.17).

In addition to varying with respect to the allelic discrimination reaction and detection method, the different SNP genotyping platforms also vary in their assay formats. The assay format can be based on a homogeneous reaction that occurs in solution or on a reaction that occurs on a solid support, such as a glass slide, a silicon chip, or a latex bead, among others. Normally, homogeneous reactions are more easily automated because they do not require separation or purification steps after the allelic discrimination reaction. However, these reactions are limited in their multiplex capacity.

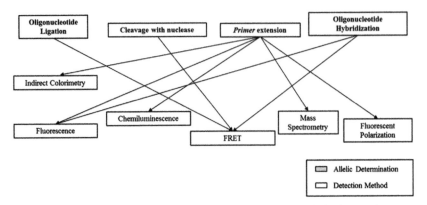

FIGURE 2.17   Detection methods that can be used for each allelic discrimination reaction.

A large number of SNP genotyping platforms are available, and many more are still undergoing development. As an example, it is possible to cite the commercial platforms GoldenGate (Illumina), Infinium (Illumina), SNPstream (Beckman Coulter), GeneChip (Affymetrix), Perlegen Wafers, and Molecular Inversion Probe – MIP (Affymetrix). The details of these platforms have been described by Syvanen (2001, 2005) and Fan et al. (2006). Generally, GoldenGate is based on bead arrays and on the principle of allele-specific primer extension; it permits the analysis of 1536 SNPs by array. The Infinium platform also uses bead arrays associated with allele-specific primer extension; however, the platform is capable of genotyping hundreds to thousands of SNPs in dozens of individuals simultaneously. SNPstream combines primer extension by minisequencing and slide array. This system allows the processing of 4600 to 3,000,000 genotypes per day; however, the number of SNPs analyzed per assay is smaller than for other platforms. The GeneChip and Perlegen Wafer systems are based on oligonucleotide hybridization. The MIP platform uses slide arrays and primer extension by minisequencing for the analysis of 10,000 SNPs per assay.

There is no ideal protocol, and many factors must be considered when choosing the most suitable platform for use in any project; these factors include sensitivity, reproducibility, accuracy, multiplex capacity for large-scale analysis, cost in terms of initial investment in equipment and per data point, and flexibility of the technique for use in other applications, as well as the time required for data analysis. Another point that should be observed is the type of project that will be carried out; in some cases, it is necessary to analyze a few SNPs in many individuals, and in other cases, analysis of many SNPs in a limited number of individuals is required. Figure 2.18 lists the numbers of SNPs and the numbers of samples that some SNP genotyping platforms are able to analyze.

SNP markers can be used in the construction of high-resolution genetic maps, in the mapping of trait of interest, in marker-assisted selection, in genetic

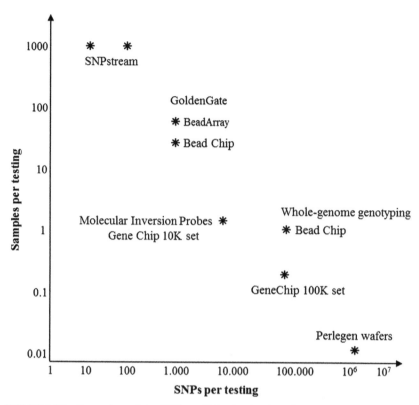

FIGURE 2.18 Comparison of the SNP's multiplex level and number of samples analyzed per assay in different SNP genotyping platforms. (*Figure adapted from Syvanen, 2005*)

studies, in the analysis of population structure, in phylogenetic analysis, and in associative mapping based on linkage disequilibrium.

CHOICE OF MOLECULAR MARKER

The choice of the most appropriate molecular marker for genetic and plant breeding studies must be made on the basis of the ease of developing a useful technique coupled with the efficiency of data evaluation, interpretation, and analysis. The chosen marker must provide easy access and availability, rapid response and high reproducibility, and allow information exchange between laboratories and between populations and/or different species; it must also permit automation of data generation and subsequent analysis. Other desirable characteristics include a highly polymorphic nature, codominant inheritance (permitting the identification of homozygous and heterozygous individuals), frequent occurrence in the genome, and neutral selection (selection free from interference by management practices and environmental conditions). In addition to the characteristics of the marker, the goals of the project, the availability of financial, structural, and personal resources, convenience, and the availability of facilities for the development of the assay, as well as the genetic trait of the species under study, should all be considered.

Of the markers described in this chapter, AFLP allows the analysis of multiple loci in a single assay due to its efficiency in wide and simultaneous sampling of the genome. The ability of AFLP to differentiate individuals within a population makes it useful for diversity studies, analysis of gene flow, and varietal discrimination (fingerprint of varieties) (Arif et al., 2010). AFLP is also a useful alternative for studies of species for which a limited number of SSR and SNP markers are available because it can use universal primers regardless of the species. However, assay and analysis of data from AFLP are complex and time consuming compared with those of other markers. In addition, AFLP is a dominant marker and requires relatively large quantities of high-quality DNA.

SSR markers are distinguished by their highly polymorphic nature; consequently, they can be used to detect a large number of alleles per locus. These markers are especially informative due to their codominant and multi-allelic nature and their high values of heterozygosity and PIC (Polymorphism Information Content). Frequent and distributed at random throughout the genome, SSRs combine informativity with a wide coverage of the genome, robustness, and reproducibility, making them one of the most desirable markers for detailed studies of genetic structure, mapping, and fingerprinting of cultivars. The great disadvantage of the SSR technique lies in the fact that it analyzes only a small sample of the genome in each assay, even when multiplex testing is used. Because SSR markers are locus specific, only a few loci are analyzed per experiment.

SNP markers offer the advantages of distribution and frequency in the genome; they also permit automated data collection and analysis. For these reasons, they are recommended for work involving wide genomic selection and studies of kindred plants with narrow genetic bases. Although SNP markers are biallelic in nature and offer lower resolution than multi-allelic SSR markers, this deficiency is balanced by the capacity to analyze a large number of loci per test (Arif et al., 2010). Another limitation of SNP markers lies in the cost of the equipment and the assay. Although large-scale genotyping techniques (*ultra-high-throughput genotyping*) allow the broad coverage of the genome in a single assay with relatively low cost per data point, the total cost of the experiment is still quite high.

Despite their recent development, DArT markers have been thoroughly tested in several species (Wenzl et al., 2004; Akbari et al., 2006; Xie et al., 2006; Mace et al., 2008; Tinker et al., 2009; Sansaloni et al., 2010; Heller-Uszynska et al., 2011). In the majority of the species in which DArT has been applied, the technique has been adapted and used in the characterization of germplasm collections and genetic diversity, the construction of high-resolution genetic maps, and the identification of QTL (Quantitative Trait Loci). Because DArT markers are widely distributed in the genome and can be used for genotyping on a large scale, the technique has great potential for use in selective genomics. Furthermore, DArT markers have shown good performance in analyses of polyploid species. This feature is very important because polyploidy can affect the accuracy of marker technologies in various ways. Techniques based on PCR, for example, can be affected by the presence of alternative annealing primer sites on chromosomes, the dilution of correct target sites or even competition with primers. Similarly, as in the case of RFLP, technologies based on hybridization can be affected by the occurrence of multiple hybridizations with the same probe. Markers whose development is based on prior knowledge of the DNA sequence, such as SNPs and SSRs, are of limited use in species for which there is little available information on the genome sequence. This limitation especially affects polyploid species, in which sequencing is technically more complicated. The results obtained with DArT in different polyploid species such as wheat (Akbari et al., 2006), hexaploid oat (Tinker et al., 2009), and sugar cane (Heller-Uszynska et al., 2011) have shown that ploidy does not affect the usefulness or reproducibility of these markers in detecting polymorphism. DArT markers, therefore, may be a good alternative for genetic studies of polyploidy.

REFERENCES

Akbari, M., Wenzl, P., Caig, V., Carling, J., Xia, L., Yang, S., et al., 2006. Diversity arrays technology (DArT) for high-throughput profiling of the hexaploid wheat genome. Theoretical and Applied Genetics 113, 1409–1420.

Arif, I.A., Bakir, M.A., Khan, H.A., Al Farhan, A.H., Al Homaidan, A.A., Bahkali, A.H., et al., 2010. A brief review of molecular techniques to assess plant diversity. International Journal of Molecular Sciences 11, 2079–2096.

Asif, M., Rahman, M., Mirza, J.I., Zafar, Y., 2008. High-resolution metaphor agarose gel electrophoresis for genotyping with microsatellite markers. Pakistan Journal of Agricultural Sciences 45 (1), 75–79.

Blears, M.J., Grandis, S.A.D., Lee, H., Trevors, T.J., 1998. Amplified fragment length polymorphism (AFLP): a review of the procedure and its applications. Journal of Industrial Microbiology and Biotechnology 21, 99–114.

Caixeta, E.T., de Oliveira, A.C.B., de Brito, G.G., Sakiyama, N.S., 2009. Tipos de marcadores moleculares [Types of molecular markers]. In: Borém, A., Caixeta, E.T. (Eds.), Molecular Markers [Marcadores moleculares], pp. 9–78.

Chalhoub, B.A., Thibault, S., Laucou, V., Rameau, C., Hoeffe, H., Cousin, R., 1997. Silver staining and recovery of AFLP amplification products on large denaturing polyacrylamide gels. BioTechniques 22, 216–218.

Cho, Y.G., Blair, M.W., Panaud, O., McCouch, S.R., 1996. Cloning and mapping of variety-specific rice genomic DNA sequences: amplified fragment length polymorphisms (AFLP) from silver-stained polyacrylamide gels. Genome 39, 373–378.

Ellegren, H., 2004. Microsatellites: simple sequences with complex evolution. Nature Reviews 5, 435–445.

Fan, J.B., Chee, M.S., Gunderson, K.L., 2006. Highly parallel genomic assays. Nature Reviews Genetics 7, 632–644.

Ferrão, L.F.V., Caixeta, E.T., de Souza, F., Maciel-Zambolim, E., Cruz, C.D., Zambolim, L., et al., 2013. Comparative study of different molecular markers for classifying and establishing genetic relationships in *Coffea canephora*. Plant Systematics and Evolution 299, 225–238.

Ferreira, M.E., Grattapaglia, D., 1998. Introdução ao uso de marcadores moleculares em análise genética [Introduction to the use of molecular markers in genetic analysis]. DF: Embrapa Genetic Resources and Biotechnology, Brasília. [Embrapa recursos genéticos e Biotecnologia]. p. 220.

Guichoux, E., Lagache, L., Wagner, S., Chaumeil, P., Léger, P., Lepais, O., et al., 2011. Current trends in microsatellite genotyping. Molecular Ecology Resources 11, 591–611.

Gupta, P.K., Rustgi, S., Mir, R.R., 2008. Array-based high-throughput DNA markers for crop improvement. Heredity 101, 5–18.

Heller-Uszynska, K., Uszynski, G., Huttner, E., Evers, M., Carlig, J., Caig, V., et al., 2011. Diversity Arrays Technology effectively reveals DNA polymorphism in a large and complex genome of sugarcane. Molecular Breeding 28, 37–55.

Jaccoud, D., Peng, K., Feinstein, D., Kilian, A., 2001. Diversity arrays: a solid state technology for sequence information-independent genotyping. Nucleic Acids Research 29, 1–7.

Kalia, K.R., Rai, M.K., Kalia, S., Singh, R., Dhawan, A.K., 2010. Microsatellite markers: an overview of the recent progress in plants. Euphytica 177, 309–334.

Liu, L., Li, Y., Li, S., Hu, N., He, Y., Pong, R., et al., 2012. Comparison of next-generation sequencing systems. Journal of Biomedicine and Biotechnology http://dx.doi.org/10.1155/2012/251364. ID 251364.

Mace, E.S., Xia, L., Jordan, D.R., Halloran, K., Parh, D.K., Huttner, E., et al., 2008. DArT markers: diversity analyses and mapping in *Sorghum bicolor*. BMC Genomics 9, 26.

Paran, I., Michelmore, R.W., 1993. Development of reliable PCR-based markers linked to downy mildew resistance genes in lettuce. Theoretical and Applied Genetics 85, 985–993.

Reddy, M.P., Sarla, N., Siddiq, E.A., 2002. Inter simple sequence repeat (ISSR) polymorphism and its application in plant breeding. Euphytica 128, 9–17.

Sansaloni, C.P., Petroli, C.D., Carling, J., Hudson, C.J., Steane, D.A., Myburg, A.A., et al., 2010. A high-density diversity arrays technology (DArT) microarray for genome-wide genotyping in *Eucalyptus*. Plant Methods 6, 16.

Savelkoul, P.H.M., Aarts, H.J.M., De Haas, J., Dijkshoorn, L., Duim, B., Otsen, M., et al., 1999. Amplified-fragment length polymorphism analysis: the state of an art. Journal of Clinical Microbiology 37, 3083–3091.

Selkoe, K.A., Toonen, R.J., 2006. Microsatellites for ecologists: a practical guide to using and evaluating microsatellite markers. Ecology Letters 9, 615–629.

Syvanen, A.C., 2001. Accessing genetic variation: genotyping single-nucleotide polymorphisms. Nature Reviews Genetics 2, 930–940.

Syvanen, A.C., 2005. Toward genome-wide SNP genotyping. Nature Genetics 37 (Suppl.), S5–S10.

Tinker, N.A., Kilian, A., Wight, C.P., Heller-Uszynska, K., Wenzl, P., Rines, H.R., et al., 2009. New DArT markers for oat provide enhanced map coverage and global germplasm characterization. BMC Genomics 10, 39.

Varshney, K.R., Graner, A., Sorrells, M.E., 2005. Genic microsatellite markers in plants: features and applications. Trends in Biotechnology 23, 48–55.

Vos, P., Hogers, R., Bleeker, M., Reijans, M., Van de Lee, T., Hornes, M., et al., 1995. AFLP: a new technique for DNA fingerprinting. Nucleic Acids Research 23, 4407–4414.

Wenzl, P., Carling, J., Kudrna, D., Jaccoud, D., Huttner, E., Kleinhofs, A., et al., 2004. Diversity arrays technology (DArT) for whole-genome profiling of barley. Proceedings of the National Academy of Sciences of United State of America 101, 9915–9920.

Xie, Y., Mcnally, K., Li, C.Y., Leung, H., Zhu, Y.Y., 2006. A high-throughput genomic tool: diversity array technology complementary for rice genotyping. Journal of Integrative Plant Biology 48, 1069–1076.

Chapter 3

Biometrics Applied to Molecular Analysis in Genetic Diversity

Cosme Damião Cruz, Caio Césio Salgado, and Leonardo Lopes Bhering

General Biology Department, University of Viçosa, Brazil

INTRODUCTION

Studies about genetic diversity have been of great importance for the purposes of genetic improvement and to evaluate the impact of human activity on biodiversity. They are equally important in the understanding of the microevolutionary and macroevolutionary mechanisms that act in the diversification of the species, involving population studies, as well as in the optimization of the conservation of genetic diversity. They are also fundamental in understanding how natural populations are structured in time and space and the effects of anthropogenic activities on this structure and, consequently, on their chances of survival and/or extinction. This information provides an aid in finding the genetic losses generated by the isolation of the populations and of the individuals, which will be reflected in future generations, allowing for the establishment of better strategies to increase and preserve species diversity and diversity within the species.

Genetic resources are established by accesses that represent the genetic variability organized in a set of different materials called germplasm. They comprise the diversity of genetic material contained in old, obsolete, traditional, and modern varieties, wild relatives of the target species, wild species and primitive lines, which can be used in the present or in the future, for food, agriculture, and other purposes.

The most important causes of loss of biodiversity and of genetic resources are: the destruction of habitats and natural communities; genetic vulnerability; genetic erosion; and genetic drift. The genetic diversity of the species is an important way to maintain the natural capacity to respond to climatic changes and all types of biotic and abiotic stress. In actuality, there exists great concern in evaluating biodiversity, because of the marked loss of genetic diversity, mostly due to the actions of man, replacing local varieties with modern varieties, hybrids, and, most recently, clones, so that large expanses of area are occupied by one or a few varieties or narrow genetic base.

Biotechnology Applied to Plant Breeding. http://dx.doi.org/10.1016/B978-0-12-418672-9.00003-9

The loss of this diversity will probably decrease the organisms' capacity to respond to environmental changes and will eliminate potentially useful biological information, as the genetic diversity of cultivated species and valuable biochemical compounds is still unknown.

In this sense, an important evaluation parameter is the fraction of intrapopulational and interpopulational components of the total genetic variability of a given species. In species with low genetic variability, the component interpopulational variation can be large, due to local adaptation or simply to divergence as a result of low genetic flux (Frankham et al., 2003).

Erosion constitutes the reduction of genetic diversity, with loss of individual genes and of particular combinations of genes (gene pools), like those manifested in locally adapted breeds. It has been reported that only 15 to 30 superior plant species have been responsible for 90% of food grown for human consumption. This situation reflects considerable loss of genetic diversity in plants, which represents genetic erosion over a long period of time. The principal cause of this genetic erosion of the crops is also the replacement of local varieties with exotic species and improved varieties. Thus, breeding programs – responsible for the development of superior varieties based on modern agriculture – are considered the major cause of genetic erosion.

Genetic drift is one of the factors that alter gene frequency in the form of dispersion, that is, it has a quantifiable magnitude, but the direction of its action is unpredictable and can either increase or decrease the frequency of a particular allele in the population. This is a stochastic process, acting on the populations, modifying the frequency of those alleles, and, consequently, contributing to the predominance of certain genotypic combinations in the populations. Its effects are manifested more frequently in populations with reduced effective sizes.

The evaluation of genetic diversity was originally carried out starting with phenotypic information related to morphologic characteristics or agronomic performance. However, the recent advances in molecular biology have opened new perspectives for research in species conservation and for the study of populational biology. With the use of molecular markers, it is possible to detect the existent variability directly in the DNA. Thus, studies about diversity are also distinguished by their goals, which may be targeted for genetic improvement, for evolutionary associations and for the conservation and management of genetic material.

Currently, the analysis of polymorphisms of DNA fragments has been one of the main tools in the study of biodiversity, allowing, as inferred from special distribution patterns of genetic diversity, to test even explicit hypotheses of historic biogeography and define priority areas for conservation. The evaluation of molecular polymorphisms in noncoding regions has provided important information about the various levels of genetic diversity, intra × interpopulational and intra × interspecific. Accordingly, the microsatellites (SSRs), repetitions in tandem of one to six nucleotides, are ideal tools, which are randomly distributed in the genome of most eukaryotic organisms, because they present codominance, a high degree of polymorphism, and are easily detected by PCR (Glowatzki-Mullis et al., 1995).

In a review paper Silva and Russo (2000) grouped the molecular techniques, applied to populational biology, into four main categories:

1. The first comprises problems related to the analysis of genetic variation within individuals, covering areas such as heteroplasmy (variation evaluated in the mitochondrial DNA), evolution of multigene families, as well as problems related to forensic medicine.

2. Included in the second cluster were problems that involve the variation within populations, such as the effect of endogamy and genetic bottlenecking in hereditary variation, consanguinity, nepotism, social structure, and reproductive success.

3. Clustered into the third category were studies directly related to genetic variation between populations, involving problems like bioinvasion and gene flow, structuring of natural stock in populations of extractive exploitation, as well as problems related to the taxonomic status of morphotypes or ecotypes.

4. Finally, grouped into the fourth category were the problems that involve the genetic variation above the species level, including the effects of the different life cycles on the process of differentiation and isolation of the species, in the hybridization, and in the establishment of hybrid zones, as well as the phylogenetic reconstruction of the taxa.

It is worth mentioning that the genetic material used in the works with molecular markers is, in the last analysis, a sample of the original population. Consequently, the variability contained within the sample will depend on the polymorphism and the genetic structure existent in the population and on the way in which the sampling was performed (Robinson, 1998). The polymorphism accessed, on the other hand, depends in large part upon the molecular technique adopted. In whatever situation, however, the basic presupposition is that the molecular markers analyzed are inheritable, reproducible, and independent (Silva and Russo, 2000).

GENETIC DIVERSITY BETWEEN ACCESSES OR WITHIN POPULATIONS

In order to study the diversity within the population or between accesses, beginning with molecular marker information, one can adopt clustering techniques or projection of dissimilarity measures techniques. In both cases, the matrix of dissimilarity between pairs of individuals through appropriate indexes becomes necessary. In these studies one considers as available the molecular information of dominant nature or codominant and multi-allelic information, as occurs in some situations in which it offers data derived from the employment of micro-satellite markers.

The information related to the dominant markers is depicted by binary code, using the code 1 to represent band occurrence and code 0 to characterize their absence. For multi-allelic codominant markers, the genotypic description is made by numerical codes informative of the alleles present. Thus considering the existence of three alleles (A_1, A_2, and A_3), there are the homozygous genotypes described by 11, 22, and 33 and the heterozygous genotypes described by 12, 13, and 23.

Cluster Analysis

In this case, the initial step is to choose the measure of distance (dissimilarity) to be used in the formation of the matrix of distance between the pairs of individuals, following which, multivariate methods capable of producing a cluster structure can be applied. The cluster analysis involves evaluating the individual capacity of allocation or discrimination, in their respective reference centers (populations), based on the variables evaluated, as well as formulate and test hypotheses about the causes of this agglomeration or dispersion. The definition of the measures or coefficients of (dis)similarity to be used is based on the type of variable under analysis.

The measures of dissimilarity can be clustered based on the nature of the molecular data from the following format:

1. for dominant markers, whose genotypic information is coded in binary form in the presence and absence of the band, commonly known as coefficients of (dis)similarity; and
2. for codominant markers, whose genotypic and allelic frequency information can be accessed, allowing the calculations of genetic (and/or geometric) and genotypic distances.

Coefficients of Similarity and Dissimilarity

For Dominant Markers

The coefficients of similarity (S), originally proposed for the study of numerical taxonomy, use this type of molecular information as binary variables, coded as 0 in the absence of a band and 1 in the presence of a band. In this situation, the coefficients of similarity between pairs of populations (be they exotic materials, strains, varieties, species, etc.) are obtained taking into account:

a: the number of coincidences of type 1-1 for each pair of populations;
b: the number of disagreements of type 1-0 for each pair of populations;
c: the number of disagreements of type 0-1 for each pair of populations; and
d: the number of coincidences of type 0-0 for each pair of populations.

As one treats the measures of similarity, it is recommended, in clustering analyses, to make use of measures of dissimilarity, defined by:

(a) $D = 1 - S$.
(b) $D = 1(k + S)$. In this expression, the inclusion of the constant k (generally $k = 1$) seeks to circumvent the problems of indetermination caused when the value of S is null. In this situation, if the value of S varies from 0 to 1, the value of D will vary from 1 to 0.5 for k equal to 1.
(c) $D = \sqrt{1 - S}$ or $\sqrt{2(1 - S)}$. This expression attributes Euclidian properties to some measures of dissimilarity, with the exception of the Yule coefficient, which is not metric (Gower and Legendre, 1986).

Another measure of dissimilarity is the Euclidian distance itself (EDI), given by:

$$EDI = \sqrt{\frac{b + c}{a + b + c + d}}$$

This measure is known as the Sokal binary distance. Naturally, all of the Euclidian coefficients are metric, while not every metric coefficient has Euclidian properties (Jackson et al., 1989).

In practice, the coefficients of similarity and dissimilarity most used in the works of genetic diversity have been simple coincidence (Sneath and Sokal, 1973; Jaccard, 1908; Nei and Li, 1979) (Table 3.1). Simple coincidence has, in its arithmetic complement, the advantage of being identical to the square of the average Euclidian distance (D^2_{ii}) and, since it is applied to the square root $\left(\sqrt{D^2_{ii'}} \right)$, it becomes its own Sokal binary distance.

It has been the consensus that, when working with exotic materials or differentiated species, at first somewhat related, the solution would be to adopt the Nei and Li coefficient. However, if the study is done within one population, or species, in which coincidences of band occurrence can be

TABLE 3.1 Expressions of the Principal Coefficients of Similarity (S)

Coefficients	Expression	Interval
Simple coincidence	$S_{CS} = \frac{a+b}{a+b+c+d}$	0, 1*
Jaccard	$S_J = \frac{a}{a+b+c}$	0, 1
Sorensen-Dice or Nei and Li	$S_{SD} = \frac{2a}{2a+b+c}$	0, 1

* Corresponds to the maximum similarity between two individuals.

admitted as an expected phenomenon, the use of the Jaccard coefficient is recommended.

For Codominant Markers

With codominant markers, such as the microsatellites, it is possible to identify heterozygous and homozygous genotypes in each site studied, generating more information on a genetic level. The similarity between pairs of accesses can be calculated by the following indices:

- *Unweighted index*

$$S_{ii'} = \frac{1}{2L} \sum_{j=1}^{L} c_j$$

in which L is the number of sites studied and c_j is the number of common alleles between the pairs of accesses i and i'.

The number of alleles common between accesses of genotypes A_iA_i and $A_i\,A_i$ (or $A_i\,A_j$ and A_jA_i) is two; between $A_i\,A_i$ and $A_i\,A_j$ (or $A_i\,A_j$ and $A_i\,A_k$), one; and between $A_i\,A_j$ and A_kA_l, zero. The specified index varies from 0 to 1.

- *Weighted index*

$$S_{ii'} = \frac{1}{2} \sum_{j=1}^{L} p_j c_j$$

in which $p_j = \frac{a_j}{A}$ is the weight associated with site j, determined by a_j: total number of alleles studied, being $\sum_{j=1}^{L} p_j = 1$ and c_j number of alleles common between the pairs of accesses i and i'. For cases in which $a_1 = a_2 \ldots = a_L = a$, has $A = a_L$, and the weighted index becomes equal to the unweighted.

- *Smouse and peakall d^2 index*

$$d_{ii'}^2 = \frac{1}{2L} \sum_{j=1}^{L} \sum_{k=1}^{a_j} \left(y_{ijk} - y_{i'jk} \right)^2$$

in which y_{ijk} refers to the quantity of alleles k in site j that individual I presents. If only the site were considered, it would be possible to obtain the distances shown in Table 3.2.

By this index, two homozygous genotypes and two heterozygous genotypes that do not share common alleles have dissimilarity equal to two, for one site. This dissimilarity would be even greater if two genotypes did not share common alleles and one of them was homozygous and the other heterozygous.

TABLE 3.2 Similarity between Pairs of Accesses by *Smouse and Peakall d^2* Index

Genotypes	Alleles				
	i	j	k	L	d^2
A_iA_i and A_iA_i	0				0
A_iA_i and A_jA_j	2	−2			2
A_iA_i and A_iA_j	1	−1			1
A_iA_i and A_jA_k	2	−1	−1		3
A_iA_j and A_iA_i	−1	1			1
A_iA_j and A_jA_j	1	−1			1
A_iA_j and A_iA_j	0	0			0
A_iA_j and A_iA_k	0	1	−1		1
A_iA_j and A_jA_k	1	0	−1		1
A_iA_j and A_kA_l	1	1	−1	−1	2

* Distance between individuals.

Clustering Methods

In many situations, the researcher is interested in evaluating the clustering patterns, formulating and testing hypotheses about the similarity of diversity obtained. However, because the number of dissimilarity estimates is relatively large (equal to the $n(n-1)/2$, in which n is the number of accesses considered in the study), the recognition of homogeneous clusters by a simple visual examination of these estimates becomes impractical. Thus, in order to carry out this task, methods of clustering or projections of distances in two-dimensional or tridimensional graphs are used, in which each coordinate is obtained beginning with the chosen measure of dissimilarity.

The versatility and the discriminatory power of the clustering methods have permitted their application in the more varied areas of the science (Everitt, 1993). The literature presents innumerable clustering techniques, which distinguish themselves by the type of result to be provided and by different ways to define the proximity between an individual and a preformed cluster between any two clusters. In most cases, one does not know, *a priori*, the number of clusters to be established, and different methods generate different results. The clustering analysis is one purely exploratory technique, which seeks the generation of hypotheses about the established agglomeration pattern and can be supplemented or complemented by other visualization techniques (Dias, 1998). Otherwise no hypothesis about the probability distribution of the data is necessary.

Three clustering approaches in particular stand out. The first of these relates to the techniques that produce dendrograms, in which the first step is to calculate the measures of dissimilarity between all of the possible pairs of accesses and, thus, form clusters by agglomerative or divisive processes; the second involves partitions in a cluster in which the individuals can move themselves to the outside or to the inside of this and other clusters in different stages of the analysis. Lastly, there are techniques based on graphic dispersion in which one considers the relative position of accesses in two-dimensional or tridimensional graphs.

Clustering Method of Optimization

The clustering method of optimization most commonly used is the Tocher method, which requires one to obtain the matrix of dissimilarity, from which the most similar pair of individuals is identified. These individuals form the initial cluster. Beginning from there, the possibility of inclusion of other individuals is evaluated; adopting the criteria of the average intracluster distance should be lower than the average intercluster distance.

The entrance of an individual into a cluster always increases the average value of distance within a cluster. Thus, one can make the decision to include the individual in a cluster through the comparison between the increase in average value of distance within the cluster and the maximum level permitted, which can be established arbitrarily or adopted, as has generally been done, the maximum value (θ) of the measure of dissimilarity encountered in the set of the smallest distances involving each individual.

In this case, the distance between the individual k and the cluster formed by the individuals ij is given by $d_{(ij)k} = d_{ik} + d_{jk}$.

During the clustering process, it is necessary to evaluate the increase in the total of the diversity of the cluster, in stage t, by the inclusion of one access k of larger similarity to the cluster. Thus the values shown in Table 3.3 should be observed.

TABLE 3.3 Increase in the Total of Diversity during the Clustering Process

Stage	Number of Accesses	Number of Distances	Total of the Distances
T	n	$n(n-1)/2$	d (cluster)
$t+1$	$n' = n+1$	$(n+1)n/2$	d (cluster$+k$) $= d$(cluster)$+ d$ (cluster)k
Increase		n	d (cluster)k

The inclusion, or not, of the individual k in the cluster is then done considering that the average increase promoted by the inclusion of one individual k in a previously established cluster will be less than θ.

If

$$\frac{d_{(grupo)k}}{n} \leq \theta$$

includes the individual k in the cluster and if

$$\frac{d_{(grupo)k}}{n} > \theta,$$

the individual k is not included in the cluster, n being the number of individuals that constitute the original cluster.

Sequential, Agglomerative, Hierarchical, Nonoverlapping Clustering Methods

In these methods, also known as SAHN, one must recalculate the coefficient of dissimilarity between the established clusters and the possible candidates for future admission into the cluster every clustering stage. Other than this, one reconsiders the admission criteria for new individuals to previously established clusters (Sneath and Sokal, 1973).

Such methods are processed by a successive series of fusions. The n individuals are classified in $n-1$, $n-2$, ..., until one cluster, in turn, based on hierarchical subdivisions operated by the distance matrix, permitting the generation of a two-dimensional classification tree, called a dendrogram. In all of the SAHN methods, at every stage, the clustering algorithm is based on the lowest distance among all of the clusters and/or pairs of individuals like the criteria for the formation of a new cluster.

It is worth noting that the choice of a clustering method depends on the material and goals in question, as different clustering methods can be influenced by the chosen coefficient of dissimilarity (Jackson et al., 1989).

Method of Average Binding among Clusters or UPGMA (Unweighted Pair-Cluster Method using Arithmetic Averages)

The method of unweighted average binding among clusters, better known as UPGMA, has been used most frequently in ecology and systematics (James and McCulloch, 1990) and in numerical taxonomy (Sneath and Sokal, 1973). UPGMA is treated as a clustering technique that uses the (unweighted) arithmetic averages of the measures of dissimilarity, thus avoiding characterizing the dissimilarity by extreme values (minimum and maximum) between the considered genotypes.

As a general rule, the construction of the dendrogram is established by the genotype of greatest similarity. However, the distance between an individual k and a cluster, formed by individuals i and j, and supplied by:

$$d_{(ij)k} = \text{average } (d_{jk}; d_{jk}) = \frac{d_{ik} + d_{jk}}{2}$$

that is, $d_{(ij)k}$, given by the average of the set of the distances of the pairs of individuals (i and k) and (j and k).

The distance between two clusters is given by:

$$d_{(ij)(klm)} = \text{average } (d_{lk}; d_{il}; d_{im} d_{jk}; d_{jl}; d_{jm}) = \frac{d_{ik} + d_{il} + d_{im} + d_{jk} + d_{jl} + d_{jm}}{6}$$

that is, the distance between two clusters, formed respectively by individuals (i and j) and (k, l, and m), and determined by the average between the elements of the set, whose elements are distance between pairs of individuals of clusters (i and k), (i and l), (i and m), (j and k), (j and l), and (j and m).

A general expression for the unweighted average among clusters can be presented in the following manner:

$$d_{(ij)k} = \frac{n_i}{n_i + n_j} d_{ik} + \frac{n_j}{n_i + n_j} d_{jk}$$

in which $d_{(ij)k}$ is defined as the distance between the cluster (ij), with an internal size n_i and n_j, respectively, and the cluster k. In this expression, the indexers i, j, and k are characterized as individuals or clusters. This interpretation should be the same for the subsequent methods.

Thus for the calculation of the distance $d_{(12)3}$, in which one considers the cluster formed by the accesses 1 ($n_i = 1$) and 2 ($n_j = 1$), one has:

$$d_{(12)3} = \frac{1}{1+1} d_{13} + \frac{1}{1+1} d_{23} = \frac{d_{13} + d_{23}}{2}$$

For the calculation of the distance $d_{(12.3)4}$, in which one considers the cluster formed by the accesses 1 and 2 ($n_i = 2$) and 3 ($n_j = 1$), one has:

$$d_{(12.3)4} = \frac{2}{2+1} d_{(12)4} + \frac{1}{2+1} d_{34} = \frac{2}{3} \left(\frac{1}{1+1} d_{14} + \frac{1}{1+1} d_{24} \right)$$
$$+ \frac{d_{34}}{3} = \frac{d_{14} + d_{24} + d_{34}}{3}$$

Method of Weighted Average Binding among Clusters or WPGMA (Weighted Pair-cluster Method using Arithmetic Averages)

In this method, at each step, one obtains a weighted average of the new coefficients of distance, which will be recalculated to compose the "new" matrix of distance.

The expression for the calculation of the weighted average is given by:

$$d_{(ij)k} = \frac{d_{ik} + d_{jk}}{2}$$

Thus, to calculate the distance $d_{(12)3}$, one has:

$$d_{(12)3} = \frac{d_{13} + d_{23}}{2}$$

To calculate the distance $d_{(12.3)4}$, in which one considers the cluster formed by accesses 1 and 2 with the junction of 3, one has

$$d_{(12.3)4} = \frac{d_{(12)4} + d_{34}}{2} = \frac{\frac{d_{14}+d_{24}}{2} + d_{34}}{2} = \frac{d_{14} + d_{24} + 2d_{34}}{2}$$

- *Ward's minimum variance method*
  In this method, one considers, for the initial formation of the cluster, those individuals that provide the smallest sum of squares of the deviations. One admits that, in whatever stage, there is loss of information due to clustering performed, which can be quantified by the ratio between the sum of squares of the deviations within the cluster in formation and the total sum of squares of the deviations. The sum of squares of the deviations within the cluster is calculated considering only the accesses within the cluster in formation, and the sum of squares of the total sum of squares of the deviations is calculated considering all of the individuals available for clustering analysis.
  The clustering is made beginning with the sums of squares of the deviations among accesses or, alternatively, beginning with the square of the Euclidian distance, since one verifies relationship:

$$SQD_{ii'} = \frac{1}{2}d_{ii'}^2$$

in which $SQD_{ii'} = \sum_{j=1}^{v} SQD_{j(ii')}$, $SQD_{j(ii')}$ being the sum of squares of the deviations, for which the jth variable, considering the accesses i and i'; and $d_{ii'}^2 = \sum_{j=1}^{v} (X_{ij} - X_{i'j})^2$ in which $d_{ii'}^2$ is the square of the Euclidian distance between the genotype si and i'; v is the number of characteristics evaluated; e X_{ij}: value of the character j for the genotype i.

The total sum of squares of the deviations is given by $SQDTotal = \frac{1}{g}\sum_{i<}^{g}\sum_{i'}^{g} d_{ii'}^2$, g being the number of accesses to be clustered.

In this clustering analysis, one identifies in matrix D (whose elements are the squares of the Euclidian distances – $d_{ii'}^2$) or in matrix S (whose elements are the sums of the squares of the deviations – $SQD_{ii'}$) the pair of accesses that provide the smallest sum of squares of the deviations. With these accesses clustered, a new matrix of dissimilarity, of inferior dimension, is recalculated, considering that:

$$SQD_{(ijk)} = \frac{1}{k}d_{(ijk)}^2$$

(k is the number of accesses in the cluster, which, in this case, is equal to 3).

$$d^2_{(ijk)} = d^2_{(ij)} + d^2_{(ij)k} = d^2_{ij} + d^2_{ik} + d^2_{jk}$$

and whereas

$$SQD_{(ijkm)} = \frac{1}{k}d^2_{(ijkm)}$$

(k is the number of accesses in the cluster, which, in this case, is equal to 4)

$$d^2_{(ijkm)} = d^2_{ij} + d^2_{ik} + d^2_{jk} + d^2_{im} + d^2_{jm} + d^2_{km}$$

and thus successively.

In the procedure, one carries out the clustering analysis providing the $g-1$ clustering steps for which the dendrogram will be formed.

Clustering Methods based on Graphic Dispersion

The clustering methods can involve higher-order matrices of dissimilarity, and the clustering performed may provoke loss of information of the degree of dissimilarity, principally of the individuals belonging to the same cluster. One alternative to the clustering methods, for the evaluation of diversity among genotypes, is the analysis of graphic dispersion, normally using the two-dimensional or tridimensional space.

It is recommended that one uses the graphic analysis when variables can be estimated – to be used in Cartesian axes – which involve the maximum available variation in the original data. In this graph, when the degree of distortion between the original distances and those represented is minimal, one can infer the dissimilarity between all of the genotypes considered, with relative ease. However, be careful to delineate the clusters based on the clustering analysis, avoiding subjectivity, since there may be an optimization criterion.

Projection is recommended when the interest is to graphically represent distances obtained by any coefficient, chosen by the researcher by describing the phenomenon studied within the biological principles of interest. The possibility of graphically representing any measure between pairs of genotypes makes this a more flexible methodology and one of great interest, principally in discrete binary or multicategoric data, in which different coefficients of similarity are adopted.

2D Projection

In this procedure, the measures of dissimilarity are converted into scores relative to two variables (X and Y), which, when represented as dispersion graphs, will reflect, in the two-dimensional space, the distances originally obtained starting with the v-dimensional space (v = number of characters used to obtain the distances).

The viability of the use of this technique is evaluated by the correlation between the original distances and those that will be represented on the dispersion graph, or by the degree distortion $(1 - \alpha)$, considering that:

$$\alpha = \frac{\sum_{i<} \sum_{i'} d_{gii'}^2}{\sum_{i<} \sum_{i'} d_{oii'}^2}$$

in which $d_{gii'}^2 \, e \, d_{oii'}^2$ are the graphic distances (two-dimensional space) and original distances (v-dimensional), respectively, of all of the pairs of individuals i and i'.

The value of the coefficient of stress (s) has also been used as a measure of the adequacy of the graphic representation, given by:

$$s = 100 \sqrt{\frac{\sum_{i<} \sum_{i'} \left(d_{gii'} - d_{oii'} \right)^2}{\sum_{i<} \sum_{i'} d_{oii'}^2}}$$

Generally, one accepts the graphic representation as satisfactory when the value of the correlation between the original measures of distance and graphic measures of distance are above 0.9 and the values of distortion and stress are below 20%.

To make the representation of the measures of similarity on graphs, one calculates the coordinate of the most divergent measures and, thereafter, those that show, in descending order, the greatest diversities with the genotypes already considered, as described by Cruz and Viana (1994). I and j being the most divergent genotypes, one arbitrarily considers that the coordinate i will equal (0,0), and that coordinate j will equal (d_{ij}, 0).

The next genotype k to be considered for graphic representation will be the one with the highest $d_{(ij)k}$ value, provided by:

$$d_{(ij)k} = d_{ik} + d_{jk}$$

Note that these values ($d_{(ij)k}$) are used only to rank the genotypes in relation to the diversity with those whose coordinates have already been established. However, the coordinate of this third genotype, given by (X_k, Y_k), is established mathematically, considering the properties of a triangle. Thus one has:

$$X_k = \frac{d_{jk}^2 - d_{ik}^2 - d_{ij}^2}{-2d_{ij}} \quad \text{and} \quad Y_k = \sqrt{d_{ik}^2 - X_k^2}$$

The same criterion is used for the next unit ℓ, that is, one chooses ℓ such that the value of $d_{(ijk)\ell}$ is greatest among all. Thus, one has:

$$d_{(ijk)\ell} = d_{i\ell} + d_{j\ell} + d_{k\ell}$$

The coordinate of the rest of the units is estimated statistically, seeking to minimize the distortion between the original distance and the graphic distance. Thus, the coordinate of unit ℓ is estimated considering that:

(a) genotype i presents coordinate $(0,0)$.
(b) genotype j presents coordinate (X_j, Y_j) in which $X_j = d_{ij}$ and $Y_j = 0$.
(c) genotype k presents (X_k, Y_k) X_k and Y_k being estimated according to the mathematic expression previously given.

The genotype ℓ presents coordinates (X_ℓ, Y_ℓ) estimated by the system of equations:

$$d_{\ell i}^2 = X_\ell^2 + Y_\ell^2$$
$$d_{\ell j}^2 = (X_j - X_\ell)^2 + (Y_j - Y_\ell)^2 = X_j^2 - 2X_j X_\ell + X_\ell^2 + Y_j^2 - 2Y_j Y_\ell + Y_\ell^2$$
$$d_{\ell k}^2 = (X_k - X_\ell)^2 + (Y_k - Y_\ell)^2 = X_k^2 - 2X_k X_\ell + X_\ell^2 + Y_k^2 - 2Y_k Y_\ell + Y_\ell^2$$

This system can be placed beneath the matricial equation $Y = X\beta + \varepsilon$ to obtain:

$$Y = \begin{bmatrix} d_{\ell j}^2 - d_{\ell i}^2 - d_{ij}^2 \\ d_{\ell k}^2 - d_{\ell i}^2 - d_{ik}^2 \end{bmatrix}; \quad X = -2 \begin{bmatrix} X_j & Y_j \\ X_k & Y_k \end{bmatrix}; \quad \beta = \begin{bmatrix} X_\ell \\ Y_\ell \end{bmatrix} e \varepsilon = \begin{bmatrix} \varepsilon_1 \\ \varepsilon_2 \end{bmatrix}$$

for the rest of the coordinates, lines in vector Y and in matrix X, which will then have the dimensions $(m-2) \times 1$ and $(m-2) \times 2$, respectively, m being the number of genotypes hitherto studied.

The solution of the system obtained by $X'X\hat{\beta} = X'Y$, so that the coordinate estimated by the genotype (symbol) presents the least distortion of distance with the others, whose coordinates have already been established. Thus consider:

$$X'X = 4 \begin{bmatrix} \sum_{n=1}^{m-2} X_n^2 & \sum_{n=1}^{m-2} X_n Y_n \\ \sum_{n=1}^{m-2} X_n Y_n & \sum_{n=1}^{m-2} Y_n^2 \end{bmatrix} \text{and } X'Y = -2 \begin{bmatrix} \sum_{n}^{m-2} \Delta_{\ell n} X_n \\ \sum_{n}^{m-2} \Delta_{\ell n} Y_n \end{bmatrix}$$

being:

$$\Delta_{\ell j} = d_{\ell j}^2 - d_{\ell i}^2 - d_{ij}^2$$
$$\Delta_{\ell k} = d_{\ell k}^2 - d_{\ell i}^2 - d_{ik}^2$$
$$\cdots$$
$$\Delta_{\ell n} = d_{\ell n}^2 - d_{\ell i}^2 - d_{in}^2$$

Note the fact that n is an indexer that assumes values correspondent to the points (genotypes) whose coordinates have already been calculated. Thus, $n = i$, j, k, l. The minimization of the distortions allows one to obtain the following solution:

$$X_\ell = \frac{\left(\sum X_n Y_n\right)\left(\sum Y_n \Delta_{\ell n}\right) - \left(\sum Y_n^2\right)\left(\sum X_n \Delta_{\ell n}\right)}{2\left[\sum X_n^2 \sum Y_n^2 - \left(\sum X_n Y_n\right)^2\right]}$$

and

$$Y_\ell = \frac{\left(\sum X_n Y_n\right)\left(\sum X_n \Delta_{\ell n}\right) - \left(\sum X_n^2\right)\left(\sum Y_n \Delta_{\ell n}\right)}{2\left[\sum X_n^2 \sum Y_n^2 - \left(\sum X_n Y_n\right)^2\right]}$$

Application Example

It will be considered, like the illustration, the projection of the measures of dissimilarity among five genotypes expressed by the arithmetic complement of the Jaccard index, given by:

$$D = \begin{bmatrix} 0 & 0.88 & 0.50 & 0.33 & 0.67 \\ & 0 & 0.88 & 0.83 & 0.50 \\ & & 0 & 0.75 & 0.50 \\ & & & 0 & 0.89 \\ & & & & 0 \end{bmatrix}$$

In order to make the projection of these measures of dissimilarity on the two-dimensional space, the following steps should be adopted:

Step 1. Identification, in the matrix of dissimilarity, of the most divergent genotypes, which in the case considered are 4 and 5, whose distance is equal to 0.89. The following coordinates are established:

Genotype 5: coordinate (0,0)
Genotype 4: coordinate $(d_{45};0) = (0.89;0)$

Step 2. Establishment of the order of dissimilarity in relation to genotypes 4 and 5, through:

$$d_{(45)1} = d_{14} + d_{15} = 0.33 + 0.67 = 1.00$$
$$d_{(45)2} = d_{24} + d_{25} = 0.83 + 0.50 = 1.33$$
$$d_{(45)3} = d_{34} + d_{35} = 0.75 + 0.50 = 1.25$$

One therefore identifies genotype 2 as having the greatest distance ($d = 1.33$) in relation to the others, whose coordinates have already been established. One obtains the coordinate of this genotype through these equations:

$$X_k = X_2 = \frac{d_{jk}^2 - d_{ik}^2 - d_{ij}^2}{-2d_{ij}} = \frac{d_{24}^2 - d_{25}^2 - d_{45}^2}{-2d_{45}} = \frac{0.83^2 - 0.50^2 - 0.89^2}{-2(0.89)} = 0.1984$$

and

$$Y_k = Y_2 = \sqrt{d_{ik}^2 - X_k^2} = \sqrt{d_{25}^2 - X_2^2} = \sqrt{0.50^2 - 0.1984^2} = 0.4589$$

Step 3. Establishment of the order of dissimilarity in relation to genotypes 2, 4, and 5, through:

$$d_{(245)1} = d_{12} + d_{14} + d_{15} = 0.88 + 0.33 + 0.67 = 1.88$$
$$d_{(245)3} = d_{23} + d_{34} + d_{35} = 0.88 + 0.75 + 0.50 = 2.13$$

Therefore one identifies genotype 3 as having the greatest distance $(d = 2.13)$ in relation to the others whose coordinates have already been established. One obtains the coordinate of this genotype through the information shown in Table 3.4.

In this case, $i = 5$ (i is the genotype with the coordinate $(0,0)$ and $d_{\ell i}^2 = d_{35}^2 = 0.50^2 = 0.25$.
Thus:

$$X_\ell = \frac{\left(\sum X_n Y_n\right)\left(\sum Y_n \Delta_{\ell n}\right)\left(\sum Y_n^2\right)\left(\sum X_n \Delta_{\ell n}\right)}{2\left[\sum X_n^2 \sum Y_n^2 - \left(\sum X_n Y_n\right)^2\right]}$$

$$X_3 = \frac{(0.0910)(0.1259) - (0.2106)(-0.3736)}{2[(0.8315)(0.2106) - (0.0910)^2]} = 0.2702$$

and

$$Y_\ell = \frac{\left(\sum X_n Y_n\right)\left(\sum X_n \Delta_{\ell n}\right) - \left(\sum X_n^2\right)\left(\sum Y_n \Delta_{\ell n}\right)}{2\left[\sum X_n^2 \sum Y_n^2 - \left(\sum X_n Y_n\right)^2\right]}$$

$$Y_3 = \frac{(0.0910)(-0.3736) - (0.8315)(0.1259)}{2[(0.8315)(0.2106) - (0.0910)^2]} = -0.4157$$

TABLE 3.4 Example to Obtain the Coordinate of the Genotype by 2D Projection

n	X_n	Y_n	X_n^2	Y_n^2	$X_n Y_n$	$d_{\ell n}^2$	d_{in}^2	$\Delta_{\ell n}$	$X_n \Delta_{\ell n}$	$Y_n \Delta_{\ell n}$
4	0.8900	0.0	0.7921	0.0	0.0	0.5625	0.7921	−0.4796	−0.4268	0.0
2	0.1984	0.4589	0.0394	0.2106	0.0910	0.7744	0.2500	0.2744	0.0532	0.1259
Total			0.8315	0.2106	0.0910				−0.3736	0.1259

TABLE 3.5 Example to Obtain the Coordinate of the Genotype by 2D Projection

N	X_n	Y_n	X_n^2	Y_n^2	X_nY_n	$d_{\ell n}^2$	d_{in}^2	$\Delta_{\ell n}$	$X_n\Delta_{\ell n}$	$Y_n\Delta_{\ell n}$
4	0.8900	0.0	0.7921	0.0	0.0	0.1089	0.7921	−0.1321	−1.0076	0.0
2	0.1940	0.4589	0.0394	0.2106	0.0910	0.7744	0.2500	0.0755	0.0146	0.0346
3	0.2702	−0.4157	0.0730	0.1728	−0.1123	0.2500	0.2500	−0.4489	−0.1213	0.1866
Total			0.9045	0.3834	−0.0213				−1.1143	0.2212

Step 4. Finally, one calculates the coordinate of genotype 1, which as of yet has not been considered. One obtains the coordinate of this genotype through the information shown in Table 3.5.

In this case, $i=5$ (i and the genotype of coordinate $(0,0)$) and $d_{\ell i}^2 = d_{15}^2 = 0.67^2 = 0.4489$.

Thus:

$$X_\ell = \frac{(\sum X_nY_n)(\sum Y_n\Delta_{\ell n}) - \left(\sum Y_n^2\right)(\sum X_n\Delta_{\ell n})}{2\left[\sum X_n^2 \sum Y_n^2 - (\sum X_nY_n)^2\right]}$$

$$X_1 = \frac{(-0.0213)(0.2212) - (0.3834)(-1.1143)}{2[(0.9045)(0.3834) - (-0.0213)^2]} = 0.6099$$

and

$$Y_\ell = \frac{(\sum X_nY_n)(\sum X_n\Delta_{\ell n}) - (\sum X_n^2)(\sum Y_n\Delta_{\ell n})}{2\left[\sum X_n^2 \sum Y_n^2 - (\sum X_nY_n)^2\right]}$$

$$Y_1 = \frac{(-0.0213)(-1.1143) - (0.9045)(0.2212)}{2\left[(0.9045)(0.3834) - (-0.0213)^2\right]} = -0.2545$$

To evaluate the adequacy of the projection performed, one obtains the graphic distances starting with the following coordinates:

Genotype	1	2	3	4	5
X_i	0.6099	0.1984	0.2702	0.89	0.0
Y_i	−0.2545	0.4589	−0.4157	0.00	0.0

In this way, one has the matrix of original distances, given by the arithmetic complement of the Jaccard index (matrix D), and the new matrix obtained by graphic distances (matrix D_g), that is:

$$
D = \begin{bmatrix}
0 & 0.88 & 0.50 & 0.33 & 0.67 \\
  & 0 & 0.88 & 0.83 & 0.50 \\
  &   & 0 & 0.75 & 0.50 \\
  &   &   & 0 & 0.89 \\
  &   &   &   & 0
\end{bmatrix}
$$

See that the estimates of distances among the first three accesses (4, 5, and 2) considered in the projection process are reproduced exactly like the distances in the original matrix. To calculate the distortion, consider:

$$
\alpha = \frac{\sum_{i<} \sum_{i'} d_{gii'}^2}{\sum_{i<} \sum_{i'} d_{oii'}^2} = 97.77\%
$$

therefore being the estimate of the degree of distortion given by:

$$
\text{distortion degree} = 1 - \alpha = 100 - 97.77 = 2.23\%
$$

In this example, the correlation among the original distances and the estimates was 0.9750, being significant at 1% of probability by the t-test made with $g(g-1)/2$ degrees of freedom, in which g is the dimension of the matrix of dissimilarity.

The estimate of the coefficient of stress is given by

$$
s = 100 \sqrt{\frac{\sum_{i<} \sum_{i'} (d_{oij} - d_{gij})}{\sum_{i<} \sum_{i'} d_{oii}^2}} = 6.54\%
$$

The process of dispersion of the measures of dissimilarity in the plane can be considered satisfactory when the coefficients that express the degree of distortion are lower than 20%. The result is presented graphically, as illustrated in Figure 3.1.

Through clustering techniques that pointed to the formation of four clusters, genotype 5, defined as an isolated cluster in the clustering methods, is closest to Cluster III, formed by genotypes 1, 2, 3, and 4.

3D Projection

In this type of projection, the measures of dissimilarity are converted into scores relative to three variables (X, Y, and Z), which, when represented in dispersion

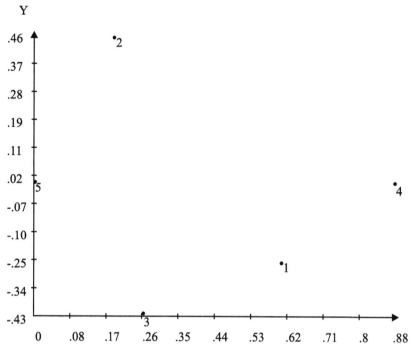

FIGURE 3.1   Projection of the dissimilarity among five genotypes, expressed by the arithmetic complement of the Jaccard index, in the two-dimensional space.

graphs, will reflect, in the tridimensional (3D) space, the distances originally obtained starting from the n-dimensional space ($n =$ number of characters used to obtain distances).

The procedure consists of calculating the coordinates of the most divergent measures and of those that show, in descending order, greatest diversity with the points (genotypes) already considered.

I and j being the most divergent units, the next unit k to be considered will be that with the highest $d_{(ij)k}$ value given by:

$$d_{(ij)k} = d_{ik} + d_{jk}$$

The same criteria is used for the next unit ℓ, that is, one chooses ℓ such that the value of $(d_{(ijk)\ell})$ is greatest among all. Thus, one has:

$$d_{(ijk)\ell} = d_{i\ell} + d_{j\ell} + d_{k\ell}$$

The coordinate of the first two units is established arbitrarily. Considering two units i and j, one has the coordinate i equal to $(0,0,0)$ and j equal to $(d_{ij}, 0,0)$. The coordinate of the third unit, given by $(X_k, Y_k, 0)$, is established mathematically, taking into account the properties of a triangle, obtaining:

$$X_k = \frac{d_{jk}^2 - d_{ik}^2 - d_{ij}^2}{-2d_{ij}}$$

and

$$Y_k = \sqrt{d_{ik}^2 - X_k^2}$$

The coordinate of the fourth unit, provided by (X_1, Y_1, Z_1), is also established mathematically through the expression:

$$X_l = \frac{d_{jl}^2 - d_{il}^2 - d_{ij}^2}{-2d_{ij}}$$

$$Y_l = \frac{d_{kl}^2 - d_{il}^2 - d_{ik}^2 + 2X_lX_k}{-2d_{ij}}$$

$$Z_l = \sqrt{d_{il}^2 - X_l^2 - Y_l^2}$$

The coordinate of the other units is estimated statistically, seeking to minimize the distortion between the original distance and the graphic distance. Thus, the coordinate of unit ℓ is estimated considering that:

Genotype i presents coordinate $(0,0,0)$.
Genotype j presents coordinate (X_j, Y_j, Z_j), in which $X_j = d_{ij}$, $Y_j = 0$, and $Z_j = 0$.
Genotype k presents coordinate (X_k, Y_k, Z_k), $Z_k = 0$ and X_k and Y_k being estimates according to the mathematical expression previously given.
Genotype l presents coordinate (X_l, Y_l, Z_l) estimated as previously described.
Genotype m presents coordinate (X_m, Y_m, Z_m) estimated by the system of equations:

$$d_{mi}^2 = X_m^2 + Y_m^2$$
$$d_{mj}^2 = (X_m - X_j)^2 + (Y_m - Y_j)^2 + (Z_m - Z_j)^2 = d_{ij}^2 + d_{im}^2 - 2X_jX_m - 2Y_jY_m - 2Z_jZ_m$$
$$d_{mk}^2 = (X_m - X_k)^2 + (Y_m - Y_k)^2 + (Z_m - Z_k)^2 = d_{ik}^2 + d_{im}^2 - 2X_kX_m - 2Y_kY_m - 2Z_kZ_m$$
$$d_{ml}^2 = (X_m - X_l)^2 + (Y_m - Y_l)^2 + (Z_m - Z_l)^2 = d_{il}^2 + d_{im}^2 - 2X_lX_m - 2Y_lY_m - 2Z_lZ_m$$

This system of equations can be placed beneath the matricial notation $Y = X\beta + \varepsilon$ obtaining:

$$Y = \begin{bmatrix} d_{mj}^2 - d_{ij}^2 - d_{im}^2 \\ d_{mk}^2 - d_{ik}^2 - d_{im}^2 \\ d_{ml}^2 - d_{il}^2 - d_{im}^2 \end{bmatrix} \quad X = -2 \begin{bmatrix} X_j & Y_j & Z_j \\ X_k & Y_k & Z_k \\ X_l & Y_l & Z_l \end{bmatrix} \quad \beta = \begin{bmatrix} X_m \\ Y_m \\ Z_m \end{bmatrix} \quad \varepsilon = \begin{bmatrix} e_m \\ e_m \\ e_m \end{bmatrix}$$

For the other coordinates, lines in vector Y and matrix X are added, which will then have the dimensions $(t-3) \times 1$ and $(t-3) \times 3$, respectively, being the number of genotypes hitherto studied.

The solution of the systems is obtained by $X'X\hat{\beta} = X'Y$, so that the coordinate estimated for genotype m presents less distortion from distance than the rest, whose coordinates were already established. Thus one considers:

$$X'X = 4 \begin{bmatrix} \sum_{n=1}^{t} X_n^2 & \sum_{n=1}^{t} X_n Y_n & \sum_{n=1}^{t} X_n Z_n \\ \sum_{n=1}^{t} X_n Y_n & \sum_{n=1}^{t} Y_n^2 & \sum_{n=1}^{t} Y_n Z_n \\ \sum_{n=1}^{t} X_n Z_n & \sum_{n=1}^{t} Y_n Z_n & \sum_{n=1}^{t} Z_n^2 \end{bmatrix} \quad \text{and} \quad X'Y = -2 \begin{bmatrix} \sum_{n=1}^{t} \Delta_{nm} X_n \\ \sum_{n=1}^{t} \Delta_{nm} Y_n \\ \sum_{n=1}^{t} \Delta_{nm} Z_n \end{bmatrix}$$

being:

$$\Delta_{jm} = d_{jm}^2 - d_{im}^2 - d_{ij}^2$$
$$\Delta_{km} = d_{km}^2 - d_{im}^2 - d_{ik}^2$$
$$\cdots \Delta_{lm} = d_{lm}^2 - d_{im}^2 - d_{il}^2$$

Note the fact that n is an indexer which assumes the values correspondent to the units whose coordinates have already been calculated. Thus $n = i, j, k$.

After obtaining the coordinates of each access, one calculates the efficiency of graphic projection, comparing the estimates of the original distances and of those that will be presented in the dispersion graph. As previously cited, the statistics used to measure the efficiency of the projection refer to the correlation among the original distances and those that will be represented in the dispersion graph, to the degree of distortion $(1 - \alpha)$, and to the value of stress (s).

Application Examples

For the example of evaluation of population 1 (Table 3.6), the values of dissimilarity among the first 10 accesses can be illustrated, expressed by the arithmetic complement of the Jaccard coefficient, as presented in Table 3.7.

Note that accesses 4 and 9 present the same band occurrence pattern in the 10 sites evaluated. Accesses 3 and 10 were the most dissimilar, with a value equal to 0.778.

Clustering Analysis

A matrix of distance involving all of the accesses could be established and submitted to the clustering analysis by some specific technique. Several clustering, hierarchical and optimization, and graphic dispersion methods could be used, as has already been described. The theoretic base of these methods was already presented and could be reviewed by the reader. Thus the clustering

TABLE 3.6 Genotypic Description

Ind.	Site 1	Site 2	Site 3	Site 4	Site 5	Site 6	Site 7	Site 8	Site 9	Site 10
1	0	0	1	1	0	0	0	1	0	0
2	0	0	0	1	0	1	0	1	0	0
3	0	1	1	1	0	0	1	1	0	0
4	0	0	1	1	0	1	0	1	0	0
5	1	0	1	0	1	1	0	1	1	1
6	0	0	1	1	1	0	0	1	0	1
7	0	0	1	1	0	1	0	1	0	1
8	0	1	1	1	1	1	1	1	0	0
9	0	0	1	1	0	1	0	1	0	0
10	0	0	1	0	1	1	0	1	1	1
11	0	1	1	1	1	0	0	1	0	1
12	0	0	0	1	1	1	0	1	1	1
13	1	1	1	1	1	1	0	0	0	0
14	0	1	1	1	1	0	0	1	0	1
15	0	1	1	0	0	1	0	1	0	1
16	0	1	0	1	1	0	0	1	0	0
17	0	0	1	1	0	0	0	1	1	1
18	0	0	1	1	0	1	0	1	0	0
19	0	0	1	1	0	1	0	1	1	0
20	0	0	0	0	0	0	0	1	0	1
21	1	1	1	1	1	0	0	1	1	0
22	0	0	1	1	0	1	0	1	0	0
23	1	0	1	1	0	0	1	1	0	1
24	0	1	1	1	0	1	0	1	0	0
25	0	0	1	1	0	1	0	1	1	1
26	0	0	1	1	0	0	0	1	0	1
27	0	0	1	1	0	0	0	1	0	1
28	0	0	1	1	0	0	0	1	1	1
29	0	0	1	1	0	0	0	1	1	1

(Continued)

TABLE 3.6 Continued

Ind.	Site 1	Site 2	Site 3	Site 4	Site 5	Site 6	Site 7	Site 8	Site 9	Site 10
30	0	1	0	1	0	0	1	1	1	0
31	0	0	1	1	1	1	0	1	0	0
32	1	0	1	1	0	1	0	1	1	0
33	0	0	0	1	0	1	0	1	1	1
34	0	0	1	1	0	1	0	1	0	0
35	0	0	0	1	0	0	0	1	1	0
36	0	0	0	1	1	1	0	1	0	1
37	1	0	0	1	0	1	0	1	0	0
38	1	1	1	1	1	1	1	1	1	0
39	0	1	0	1	0	0	0	1	0	0
40	0	0	0	1	0	1	0	1	1	0
41	0	0	0	1	0	1	0	1	0	0
42	1	0	0	1	0	0	0	1	0	1
43	0	0	0	1	1	1	0	1	1	1
44	0	0	1	1	0	0	0	1	1	0
45	0	1	0	1	1	0	0	1	0	1
46	0	0	0	1	0	1	1	1	1	0
47	0	0	0	1	0	0	0	1	0	0
48	0	0	1	1	0	1	1	1	1	0
49	0	0	1	1	0	1	0	1	1	0
50	0	0	0	1	0	0	0	1	1	1

methodologies will be applied to the molecular data as an illustration, so that the following results can be considered.

Hierarchical Methods

In this type of analysis, it is of interest to study diversity through a dendrogram in which the relationships among accesses can be evaluated through the ramifications of the tree obtained. As an illustration, the clustering of 10 accesses belonging to population 1 will be considered (Table 3.7) studied by the clustering technique UPGMA, whose result is described as shown in Figure 3.2.

TABLE 3.7 Averages of Dissimilarity (Arithmetic Complement of Jaccard Index) among 10 Accesses. Estimated Starting from 10 Molecular Markers (Sites) (Original Data available in Table 3.6)

Accesses	A	b	c	d	$D_{ii'}$	Accesses	a	B	c	d	$D_{ii'}$
1 2	2	1	1	6	0.500	4 5	3	1	4	2	0.625
1 3	3	0	2	5	0.400	4 6	3	1	2	4	0.500
1 4	3	0	1	6	0.250	4 7	4	0	1	5	0.200
1 5	2	1	5	2	0.750	4 8	4	0	3	3	0.429
1 6	3	0	2	5	0.400	4 9	4	0	0	6	0.000
1 7	3	0	2	5	0.400	4 10	3	1	3	3	0.571
1 8	3	0	4	3	0.571	5 6	4	3	1	2	0.500
1 9	3	0	1	6	0.250	5 7	4	3	1	2	0.500
1 10	2	1	4	3	0.714	5 8	4	3	3	0	0.600
2 3	2	1	3	4	0.667	5 9	3	4	1	2	0.625
2 4	3	0	1	6	0.250	5 10	6	1	0	3	0.143
2 5	2	1	5	2	0.750	6 7	4	1	1	4	0.333
2 6	2	1	3	4	0.667	6 8	4	1	3	2	0.500
2 7	3	0	2	5	0.400	6 9	3	2	1	4	0.500
2 8	3	0	4	3	0.571	6 10	4	1	2	3	0.429
2 9	3	0	1	6	0.250	7 8	4	1	3	2	0.500
2 10	2	1	4	3	0.714	7 9	4	1	0	5	0.200
3 4	3	2	1	4	0.500	7 10	4	1	2	3	0.429
3 5	2	3	5	0	0.800	8 9	4	3	0	3	0.429
3 6	3	2	2	3	0.571	8 10	4	3	2	1	0.556
3 7	3	2	2	3	0.571	9 10	3	1	3	3	0.571
3 8	5	0	2	3	0.286						
3 9	3	2	1	4	0.500						
3 10	2	3	4	1	0.778						

Note that the dendrogram obtained the existence of accesses with 100% similarity (4 and 9), being indicative of whether there are replicates within the population or very close individuals, while the analysis has been made based on only 10 sites.

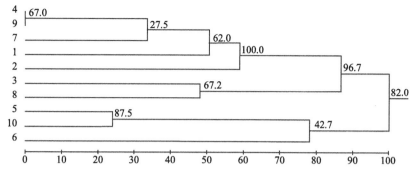

FIGURE 3.2   Results of the clustering technique UPGMA.

For a better interpretation of the results obtained by the hierarchical cluster-ing analysis, it becomes important to point out the following information:

- Value of dissimilarity obtained in the last level of fusion. In this example, the dissimilarity in the last level of fusion equaled 0.5931, inferior to the maximum verified dissimilarity between accesses 3 and 10 (which equaled 0.778).
- Identification of the cutoffs obtained via some optimal partition criterion. Using the Mojema method (1977), with a value of $k = 1.25$, cutoffs at 79 and 87% of dissimilarity are recommended.
- The consistency of nodes and bifurcations. In this type of study, it is desir-able that the consistency of the clustering pattern be evaluated; to this end, the use of the bootstrap technique is recommended. For this example, con-sider 5000 simulations, so that the consistency of the clustering of each node in the tree presented is illustrated in the figure analyzed.
- Value of the coefficient of the cophenetic correlation. For this example, the correlation between the original measures of dissimilarity and those gener-ated graphically from the dendrogram is 0.8266.

Optimization Methods

Another option for studying the existence of homogeneous clusters is through the use of optimization methods, being one of the most used as described by Tocher, cited by Rao (1952). This method, when applied to the matrix of dis-similarity expressed by arithmetic complement of the Jaccard index, provides the result presented in Table 3.8.

The result obtained by the Tocher method will be identical to that obtained by the hierarchical technique UPGMA, if a cutoff value of 50% of dissimilar-ity established in the last level of fusion is admitted for this clustering tech-nique.

TABLE 3.8 Measures of Dissimilarity among 10 Accesses. Estimated from 10 Molecular Markers (Sites)

Clusters	Accesses	Sum of the Distances	Average of the Distances
I	4,9,7, and 1	10.300	0.217
II	5 and 10	0.143	0.143
III	3 and 8	0.286	0.286
IV	6	–	–
V	2	–	–

When one uses the Tocher clustering methodology, it is important to point out

- The value of the statistic used as a global criterion for the inclusion of genotypes within a cluster. In this example, the value obtained was 0.3333.
- The values of intercluster distances, which should be inferior to the intracluster distances. In this example, the values of the distances are graphically represented via the diagram shown in Figure 3.3.

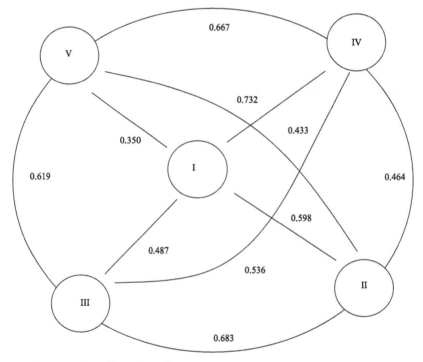

FIGURE 3.3   Values of intercluster distances.

Graphic Dispersion

It is also possible to present dissimilarity among accesses via graphic dispersion. In this case, the use of the projection of distances in 2D or 3D graphs is recommended. For the example considered, the 3D projection is as shown in Figure 3.4.

For the graphic projection, note the measures of the quality of the dispersion shown in Table 3.9.

The 3D projection presents graphic representation quality superior to that of 2D and, therefore, should be preferred. In the example in the table, there is considerable decrease in the degree of distortion and stress, as well as elevation of the correlation between the original and graphic distances for 0.986. Thus,

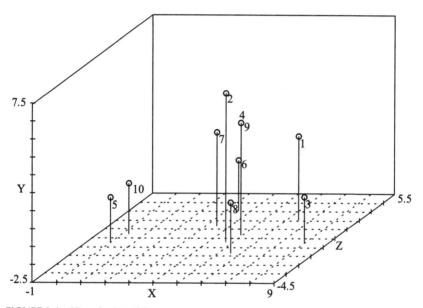

FIGURE 3.4   3D projection of distances.

TABLE 3.9 Measures of the Quality of the Dispersion for Graphic Projection

Projection	2D	3D
Distortion	14.34%	3.78%
Correlation between original and graphic distances	0.910	0.986
Stress	20.63%	6.94%

consider that the positioning of the scores of the genotypes on the graph makes it possible to appropriately infer about the genetic diversity and its clustering pattern.

DIVERSITY AND POPULATION STRUCTURE

Several methods of estimating the genetic variation and knowing the populational structure have already been developed, with varied applications on the individual, intrapopulational, and interpopulational levels.

The use of the allelic or genotypic frequencies derived from different molecular marks (sites) has been common for quantifying genetic distances to be used in the construction of phylogenetic trees for a cluster of populations or species of close relation. Besides distance, other measures, like heterozygosity, degree of fixation, and intergenotypic correlation, have been useful in the evaluation of the degree of diversity which expresses the similarity and genetic distancing within populations, provoked by genetic and environmental factors. According to Robinson (1998), among the most used measures in works with molecular markers, one points out those based on the H statistics of Nei and the F statistics of Wright, with only the first method being approached in this chapter.

Identity, Heterozygosity, and Genetic Diversity – Neigst Statistic

To calculate the genetic distances between populations it will be considered that the symbol $X_{ij\,t}$ expresses the value of the individual t, relative to site j in population i.

Being:

- $i = 1, 2...g$, in which g is the total number of populations evaluated;
- $j = 1, 2...L$, in which L is the total number of sites studied; and
- $t = 1, 2...n_i$, in which n_i is the number of individuals evaluated in population I.

$$N = \sum_{i=1}^{g} n_i$$

From the information, one estimates:

- Allelic frequency: p_{ijk}, which expresses the frequency of the allele k in site j in the population.
- Genotypic frequency: $Pi_{jmm'}$ which expresses the frequency of the genotype mm' of site j in population i.

The measure of the genetic diversity elaborated by Nei (1973, 1978) is based on genic or allelic heterozygosity (H). Its hierarchical partition permits the estimation of one component of diversity between and another component of diversity within the experimental units (populations, subpopulations, demes, etc.).

(a) Measures of Genetic Identity

The genetic identity in the ith subpopulation for a particular site is given by
$J_i = \sum_{k=1}^{a_j} p_{ijk}^2$

The genetic identity between the ith and the i'th subpopulations is provided by $J_{ii'} = \sum_{k=1}^{a_j} p_{ijk} p_{i'jk}$

From these values the following statistics can be established.

Identity between Subpopulations

The total genetic identity between all of the pairs of subpopulations for this site is provided by

$$J_{IS} = \sum_{i=1}^{g} J_i = \sum_{i=1}^{g} \sum_{k=1}^{a_j} p_{ijk}^2$$

and the average value is denoted by $\overline{J}_{IS} = \frac{J_{IS}}{g}$

Total Identity of the Subpopulations

This given by $J_{IT} = \sum_i J_i + \sum_{i \neq 1}^{g} \sum_{i'}^{g} J_{ii'} = J_{IS} + 2J_{ST}$ which is the average value provided by $\overline{J}_{IT} = \frac{J_{IT}}{g^2} = \frac{1}{g}\overline{J}_{IS} + \frac{g-1}{g}\overline{J}_{ST}$

(b) Average of Heterozygosity

For a single site in a randomly mating population, the heterozygosity is defined as $H_i = 1 - (\sum_{k=1}^{a_j} p_{ijk}^2)$ in which p_{ijk} is the frequency of kth allele for the jth site which presents a_j alleles.

For multiple sites, one calculates the average heterozygosity of a population i (H), which is the average arithmetic of the value of h for all of the sites, that is, $\overline{H}_i = \frac{1}{L} \sum_{j=1}^{L} H_{i(j)}$, in which $H_{(i)j}$ (or simply H_j) is the estimate of heterozygosity in the jth site and L is the number of sites sampled.

For a subpopulation, one has $H_i = 1 - J_i$ e $H_{ii'} = 1 - J_{ii}$.

From these values the following statistics can be established.

Heterozygosity among Subpopulations

The total genetic heterozygosity among the subpopulations for a site is given by $H_{IS} = \sum_{i=1}^{g} H_i = g - J_{IS}$ and the average value is denoted by $\overline{H}_{IS} = \frac{H_{IS}}{g} = 1 - \overline{J}_{IS}$

Heterozygosity between Subpopulations

The total genetic heterozygosity between all of the pairs of subpopulations for one site is given by $H_{ST} = \sum_{i=1}^{g} \sum_{i' < i}^{g} H_{ii'} = \frac{g(g-1)}{2} - J_{ST}$ and the average value is denoted by $\overline{H}_{ST} = \frac{2H_{ST}}{g(g-1)} = 1 - \overline{J}_{ST}$

Total Heterozygosity of the Subpopulations

The total genetic heterozygosity is provided by

$$H_{IT} = \sum_{i=1}^{g} H_i + \sum_{i=1}^{g} \sum_{i' \neq i}^{g} H_{ii'} = H_{IS} + 2H_{ST}$$

One defines the average heterozygosity of the subpopulations by the expression

$$\overline{H}_{IT} = \frac{H_{IT}}{g^2} = \frac{1}{g}\overline{H}_{IS} + \frac{g-1}{g}\overline{H}_{ST}$$

or

$$\overline{H}_{IT} = \frac{1}{g}(1 - \overline{J}_{IS}) + \frac{g-1}{g}(1 - \overline{J}_{ST}) = 1 - \overline{J}_{IT}$$

(c) Average of Diversity

The genetic diversity $(D_{ii'})$ between the ith and the i'th subpopulations and estimated by $D_{ii'} = H_{ii'} - \frac{H_i + H_{i'}}{2}$ or $D_{ii'} = \frac{J_i + J_{i'}}{2} - J_{ii'}$. As this is about diversity, one has $D_{ii} = 0$ e $D_{ii'} = D_{i'i}$

Diversity within Subpopulations

As $D_{ii} = 0$ one has $D_{IS} = \sum_i D_{ii} = 0$

Diversity between Subpopulations

$$D_{ST} = \sum_{i>} \sum_{i'} D_{ii'} = \sum_{i>} \sum_{i'} (\frac{J_i + J_{i'}}{2} - J_{ii'}) = \frac{g-1}{2} \sum_{i=1}^{g} J_i - \sum_{i=1}^{g} \sum_{i>i'}^{g} J_{ii'}$$

$$D_{ST} = \frac{g-1}{2} J_{IS} - J_{ST} \quad \overline{D}_{ST} = \frac{2}{g(g-1)} D_{ST} = \overline{J}_{IS} - \overline{J}_{ST}$$

Total Diversity of the Subpopulations

The total diversity, computed for all of the populations, is denoted by D_{IT} and calculated through:

$$D_{IT} = \sum_i D_{ii} + \sum_{i \neq} \sum_{i'} D_{ii'} = D_{IS} + 2D_{ST} = 2D_{ST}$$

The average diversity, involving all of the subpopulations, is given by:

$$\overline{D}_{IT} = \frac{D_{IT}}{g^2} = \frac{g-1}{g}\overline{J}_{IS} - \frac{g-1}{g}\overline{J}_{ST} = \overline{J}_{IS} - \left(\frac{1}{g}\overline{J}_{IS} - \frac{g-1}{g}\overline{J}_{ST}\right)$$

$$\overline{D}_{IT} = \frac{D_{IT}}{g^2} = \overline{J}_{IS} - \overline{J}_{IT}$$

or in another form

$$\overline{D}_{IT} = \frac{2D_{ST}}{g^2} = \frac{g-1}{g}\overline{D}_{ST}$$

One should remember that $\overline{J}_{IS}$ is the average genetic identity within the sub-populations, $\overline{J}_{IT}$ is the general average, and $\overline{D}_{IT}$ is the total genetic diversity of the subpopulations, including comparisons of subpopulations with themselves (only $D_{ii'}$ is null).

Briefly, one can represent the values found for the statistics as described in Table 3.10.

Differentiation between Subpopulations

The relative magnitude of the differentiation between subpopulations can be measured by:

$$G = \frac{100\overline{D}_{ST}}{\overline{H}_{ST}} \quad \text{or} \quad G_{ST} = \frac{100\overline{D}_{IT}}{\overline{H}_{IT}}$$

The measure of G_{ST} is called the coefficient of relative diversity between clusters. It varies between 0 and 1 and expresses the proportion of the total diversity explained by differences between the clusters. Mathematically, it is measured equivalent to the Wright statistic F_{ST}.

TABLE 3.10 Values Found for the Statistics

FV	Identity	Heterozygosity	Diversity
Within	$\overline{J}_{IS}$	$\overline{H}_{IS} = 1 - \overline{J}_{IS}$	$D_{IS} = 0$
Between	$\overline{J}_{ST}$	$\overline{H}_{IS} = 1 - \overline{J}_{ST}$	$\overline{D}_{ST} = \overline{J}_{IS} - \overline{J}_{ST}$ $\overline{D}_{ST} = \overline{H}_{ST} - \overline{H}_{IS}$
Total	$\overline{J}_{IT} = \frac{1}{g}\overline{J}_{IS} + \frac{g-1}{g}\overline{J}_{ST}$	$\overline{H}_{IT} = \frac{1}{g}(1 - \overline{J}_{IS}) + \frac{g-1}{g}(1 - \overline{J}_{ST})$ $\overline{H}_{IT} = 1 - \overline{J}_{IT}$	$\overline{D}_{IT} = \overline{J}_{IS} - \overline{J}_{IT}$ $\overline{D}_{IT} = \overline{H}_{IT} - \overline{H}_{IS}$

Molecular Variance Analysis – AMOVA

AMOVA is a methodology capable of studying the diversity between populations starting with molecular data and also testing a hypothesis in regards to such differentiation. One variety of molecular data, such as information of dominant or codominant markers and sequence data, can be analyzed using this method (Excoffer et al., 1992).

To analyze the diversity between populations and within them via the AMOVA methodology, one must initially calculate the matrix of dissimilarity, whose elements are the squares of the Euclidian distances between pairs of accesses, denoted by D^2. This matrix is then partitioned, clustering the information of pairs of haplotypes within each subpopulation. Thus, the original D^2 matrix can be presented in the following ways:

$$D^2 = [\tilde{T}]$$

or

$$D^2 = \begin{bmatrix} [\tilde{D}_1] & & & \\ & [\tilde{D}_2] & & \\ & & \cdots & \\ & & & [\tilde{D}_g] \end{bmatrix}$$

One has the information shown in Table 3.11.

Obtaining Sums of Squares

Sum of squares of the distances – total is given by

$$SQT = \frac{1}{2N}T$$

TABLE 3.11 Diversity between and within Populations

Subpopulations	Matrix	Total of the Elements	Number of Individuals
1	$\tilde{D}_1$	D_1	N_1
2	$\tilde{D}_2$	D_2	N_2
...	...	...	...
P	$\tilde{D}_g$	D_g	N_g
Total	$\tilde{T}$	$T \neq \sum_{i=1}^{p} D_i$	$N = \sum_{i=1}^{g} N_i$

TABLE 3.12 Outline of the Molecular Variance Analysis (AMOVA) with Data Clustered in Two Hierarchical Levels

Source of Variation	GL	SQ	QM	E(QM)
Between populations	$g-1$	SQE	QME	$\sigma_i^2 + \tilde{N}\sigma_p^2$
Within populations	$N-g$	SQD	QMD	σ_i^2
Total	$N-1$	SQT	–	σ_T^2

Sum of squares of the distances – within subpopulations is provided by

$$SQD = \sum_{i=1}^{g} \frac{D_i}{2N_i}$$

Sum of squares of the distances – between subpopulations is given by

$$SQE = SQT - SQD$$

Outline of the Variance Analysis

Having obtained the values of the sums of squares, the framework for the AMOVA analysis has been set forth. The values of the expected average squares are obtained considering a two-factor model classification, in which the averages of dissimilarity are established from the information of the variations between and within subpopulations, that is, $Y_{ij} = u + P_i + D_{ij}$ in which Y_{ij} is the average of dissimilarity between pairs of individuals (j) in a subpopulation i; u is a constant; P_i is the effect of the subpopulation i; and D_{ij} is the effect of the dissimilarity between pairs of individuals j in the subpopulation i.

The outline of the molecular variance analysis (AMOVA), with data clustered in two hierarchical levels, is presented as shown in Table 3.12.

In this case, g represents the average number of haplotypes sampled by population.

With samples of unequal sizes, $\tilde{N}$ is obtained by $\tilde{N} = \frac{N - \sum_i \frac{N_i^2}{N}}{g-1}$, where N_i is the number of individuals or haplotypes from the ith subpopulation

Estimation of Variance Components

The estimators of the variance components that express the differentiations between and within subpopulations is obtained via:

$$\hat{\sigma}_i^2 = QMD \quad \text{and} \quad \hat{\sigma}_p^2 = \frac{QME - QMD}{\tilde{N}}$$

Association between Variance Components and Statistics

The statistics obtained by AMOVA, called statistics (Φ''), reflect the correlation of the diversity of haplotypes in different levels of hierarchical subdivision (populations and individuals or haplotypes evaluated).

Thus, the coefficient of correlation between haplotypes, randomly sampled between populations, and given by:

$$r = \Phi_{ST} = \frac{Cov(Y_{ij}, Y_{ij'})}{\sqrt{V(Y_{ij})V(Y_{ij'})}}$$

Next $\Phi_{ST} = \frac{\sigma_p^2}{\sigma_T^2}$, which is an average of the relative diversity between the populations evaluated.

STATISTIC TESTS OF THE VARIANCE COMPONENTS AND STATISTICS Φ

The significance of the components σ_p^2 and Φ_{ST} can be tested by the permutation of haplotypes between populations, through the procedure based on permutation of the data. Two types of permutation are carried out. In the first, to establish the nullity of the components σ_i^2 and σ_p^2, all of the data from every piece of molecular information are permuted, independently of the populations that they belong to. In the second test, one admits as true the division of the population into subpopulation and tests only the existence variability within the subpopulations. In this case, the permutation of the data should only occur within the subpopulations.

The use of AMOVA could be extended to other hierarchized models and would serve to estimate other populational parameters.

REFERENCES

Cruz, C.D., Viana, J.M.S., 1994. A methodology of genetic divergence analysis based on sample uniprojection on two-dimensional space. Revista Brasileira de Genética, Ribeirão Preto 17 (1), 69–73.

Dias, L.A.S., 1998. Análises multidimensionais. In: Alfenas, A.C. (Ed.), Eletroforese de isoenzimas e proteínas afins. Viçosa-MG, Ed. UFV, pp. 405–475.

Everitt, B.S., 1993. Cluster Analysis. Edward Arnold, University Press, Cambridge. p. 170.

Frankham, R., Ballou, J.D., Briscoe, D.A., 2003. Introduction to Conservation Genetics. Cambridge University Press, Cambridge.

Glowatzki-Mullis, M.-L. et al., 1995. Microsatellite-based parentage control in cattle. Animal Genetics 26, 7–12.

Gower, J.C., Legendre, P., 1986. Metric and Euclidian properties of dissimilarity coefficients. Journal of Classification 3, 5–48.

Jaccard, P., 1908. Nouvelles recherches sur la distribution florale. Bulletin de la Société Vaudoise des Sciences Naturelles 44, 223–270.

Jackson, A.A., Somers, K.M., Harvey, H.H., 1989. Similarity coefficients: measures for co-occurrence and association or simply measures of occurrence. American Naturalist 133, 436–453.

James, F.C., McCulloch, C.E., 1990. Multivariate analysis in ecology and systematics: Panacea or Pandora's box? Annual Review of Ecology and Systematics 21, 129–166.

Nei, M., 1973. Analysis of gene diversity in subdivided populations. Proceedings of the National Academy of Sciences of the United States of America, Washington 70, 3321–3323.

Nei, M., 1978. Estimation of average heterozygosity and genetic distance from small number of individuals. Genetics 89, 583–590.

Nei, M., Li, W.H., 1979. Mathematical model for studying genetic variation in terms of restriction endonucleases. Proceedings of the National Academy of Sciences USA 76 (10), 5269–5273.

Rao, R.C., 1952. Advanced Statistical Methods in Biometric Research. John Wiley and Sons, New York. p. 390.

Robinson, I.P., 1998. Aloenzimas na genética de populações de plantas. In: Alfenas, A.C. (Ed.), Eletroforese de Isoenzimas e Proteínas Afins: Fundamentos e Aplicações em Plantas e Microrganismos. Viçosa, UFV, pp. 329–380.

Silva, E.P., Russo, C.A.M., 2000. Techniques and statistical data analysis in molecular population genetics. Hydrobiologia 420, 119–135.

Sneath, P.H., Sokal, R.R., 1973. Numerical Taxonomy: The Principles and Practice of Numerical Classification. W.H. Freeman, San Francisco. p. 573.

Chapter 4

Genome-Wide Association Studies (GWAS)

Marcos Deon Vilela de Resende,[a,b] Fabyano Fonseca e Silva,[b]
Márcio Fernando R. Resende Júnior,[c] and Camila Ferreira Azevedo[b]
[a]*Brazilian Enterprise for Agricultural Research on Forestry, Colombo, Brazil,* [b]*Federal University of Viçosa, Viçosa, Brazil,* [c]*University of Florida, Gainesville, FL, USA*

INTRODUCTION

Efforts to use genetic markers for genetic improvement research have diverged into two approaches: QTL (*quantitative trait loci*) identification and mapping; and marker use in genetic selection programs, through marker-assisted selection (MAS) and genome-wide selection (GWS) or genomic selection (GS). This chapter and the next cover both approaches with a special emphasis on genetic selection from genomic data.

QTL ANALYSIS AND GENOMIC SELECTION: CONCEPTS

The use of molecular genetic markers for selection and genetic improvement is based on genetic linkage between these markers and a quantitative trait locus (QTL) of interest. Thus, linkage analyses between markers and QTLs and between the proper multiple markers are essential for genetic selection from genomic information. It must be made clear that by definition, a QTL refers only to the statistical association between a genomic region and a trait.

In classical genetics, linkage between genetic factors or genes has been reported since 1906 and means that closely linked genes on a chromosome are inherited together. In other words, these genes do not segregate independently and thus they do not obey Mendel's Second Law or the Law of Independent Assortment. When these genes are close to each other on a chromosome or linkage group, the linkage is complete. When the genes are part of the same linkage group but are distant from each other, there is a partial linkage.

The calculated genetic distance between two genes is a function of the recombination frequency between the genes and forms the basis for linkage map construction. For linkage between loci to be detected and used in selection, there must be linkage disequilibrium (LD) in the studied population or family.

Biotechnology Applied to Plant Breeding. http://dx.doi.org/10.1016/B978-0-12-418672-9.00004-0

LD or gametic phase disequilibrium is a measurement of the allele interdependence at two or more loci. In a group of individuals, if two alleles from distinct loci are found together more often than would be expected, based on the product of their frequencies, it can be inferred that such alleles are in LD. Disequilibrium values near zero suggest equilibrium or independence between alleles from different genes, and values near one indicate disequilibrium or strong linkage.

LD between markers and QTLs is essential for QTL detection, MAS, and GWS. Of particular importance is the extent of this disequilibrium in a chromosome in a selected population. If a marker and a QTL are in equilibrium in the population, this marker will segregate independently from the QTL. Thus, the marker genotype of an individual has no informative value for selection. The persistence in the population of LD among linked loci depends on the distance between the loci; in other words, it depends on the recombination rate between the two loci. For closely linked loci, any LD that has been created will persist for many generations. However, for weakly linked loci (a recombination rate greater than 0.1), the LD will decrease rapidly. Although a marker (locus m) linked to a QTL (locus q) might be in linkage equilibrium in a population, there is always disequilibrium within families or crosses, even for weakly linked loci. Additionally, this disequilibrium can extend over large distances because it comes from only one recombination performed to produce the descendants of a heterozygous F_1 individual.

For example, take two linked loci, m (marker) and q (QTL), in four individuals who are heterozygous for the marker and have the following genotypes: Mq//mq, Mq//mQ, MQ//mQ, and Mq//mq. The families coming from the two first individuals will be in LD (because, for linked loci, parental gametes are more common than recombinant ones) but in opposite directions because the phase of the QTL marker differs in the two parents. The families from the two last individuals will not be in LD because the QTL does not segregate in these families. When combined across families, the four types of disequilibrium will cancel each other out, creating linkage equilibrium in the population. Thus, LD within each family is useful for QTL analysis as long as different linkage phases are considered.

In population genetics, disequilibrium generically refers to the discrepancy between the joint frequency of a combination of alleles and the product of the alleles' individual frequencies. The term normally refers to alleles from different loci in the same gamete, but can also refer to pairs of alleles of the same locus that show a lack of Hardy-Weinberg equilibrium.

QTL mapping, MAS, proposed by Lande and Thompson (1990), and GWS, proposed by Meuwissen et al. (2001), are based on the presence of linkage disequilibrium in the studied population (or cross). In this situation, marker alleles provide information about the existence and effects of loci that control quantitative traits, providing ways to estimate the effects of QTLs and allowing them to be used efficiently in genetic selection. The causes of LD in a population are mutation, migration, selection, and a small effective population size (genetic drift due to sampling). In other words, all of the factors that affect Hardy-Weinberg equilibrium in a population also affect linkage equilibrium.

Recently, molecular genetic markers that consist of SNPs (single-nucleotide polymorphisms), which are based on the detection of polymorphisms that arise from a single base change in the genome, have been used. Generally, for an SNP to be considered genetically derived, the polymorphism must occur in at least 1% of the population. SNPs are the most common form of genomic DNA variation and are preferred over other genetic markers due to their low mutation rates and ease and low cost of genotyping. Thousands of SNPs can be used to cover the entire genome of an organism with markers that are not more than 1 cM apart from each other. Microsatellite markers can also be used. These markers are efficient because they are codominant, multi-allelic, abundant, and highly transferable between individuals and species. However, the marker density of microsatellites is usually limited to a few hundred markers. This density can compromise the association study especially in populations with lower LD.

SNP markers are most often bi-allelic, as shown below:

Individual 1: TCA*C*CGCG
Individual 2: TCA*T*CGCG

In this example, there is an SNP polymorphism between the two individuals. A single base change in the DNA sequence results in a polymorphism. More than 1.5 million SNPs have been identified in the human genome. These SNPs exist at an average spacing of 2×10^{-3} cM (Hartl and Jones, 2002).

DArT (*Diversity Array Technology*) markers are also bi-allelic and well suited for GWS because they are abundant, like SNPs, and can be determined rapidly and in large numbers. However, these markers are dominant, which may be a disadvantage when compared to the codominant SNPs. GWS or GS (see Chapter 5) was proposed by Meuwissen et al. (2001) as a way to accelerate and increase the effectiveness of genetic improvement programs. GWS emphasizing the simultaneous prediction (without using significance tests for individual markers) of the genetic effects of thousands of DNA genetic markers that are spread throughout the genome of an organism, in order to the capture the effects of all loci (both small and large effects) and explain all genetic variation of a quantitative trait. To accomplish this, population level LD between the marker alleles and the genes that control the trait is essential.

Linkage Analysis (LA) and Linkage Disequilibrium Analysis (LDA)

The quantity of heritable genetic material in an individual is finite and depends on the genome size. In humans, the genome comprises approximately 35 thousand genes (Ewing and Green, 2000). Thus, a finite number of genes should control a given quantitative trait, making it possible to analyze all of the loci linked to the genetic control of the trait.

There are three basic approaches to QTL discovery: (1) candidate gene approaches, (2) mapping by linkage analysis (LA), and (3) mapping by LD analysis (LDA). The candidate gene strategy assumes that a gene involved in the underlying physiology of the trait contains a mutation that causes trait variation. This gene is then sequenced in various individuals and differences in the DNA sequences are analyzed for their associations with trait phenotype variations (Anderson and Georges, 2004). This approach has the following problems: there are a large number of potential gene candidates and it is possible that a causative mutation exists in a gene that was not chosen *a priori* as a candidate.

Mapping approaches seek to identify chromosomal regions associated with phenotypic variations in the trait of interest and do not assume *a priori* that the genes themselves are known. Such approaches are therefore based on associations between genetic marker alleles and differences in quantitative traits. A molecular DNA marker is an identifiable physical location on the chromosome for which inheritance can be monitored. The approach therefore consists of identifying markers that are associated with the trait. These markers can be the mutation affecting the quantitative trait, or it can be in LD with the cause of that effect.

A marker is considered to be informative when one can accurately determine which parental allele was transmitted to the progeny. Thus, if a genotyped parent is homozygous for the marker, this marker will not be informative in any of the progeny because it will be impossible to determine which parental allele was transmitted. If both the parent and the progeny are heterozygous, the marker might also be uninformative. If only one parent is genotyped and the progeny have the same genotype as the parent, the progeny could have received the allele from either parent. The expected frequency of individuals for whom the allele origin can be determined will be $1 - (p + q)/2$, where p and q are the frequencies of the two parental alleles for the marker. Thus, if only two marker alleles are present in the population, half of the offspring will have the same genotype as their parent. For multi-allelic loci such as microsatellites, $(p + q)$ can be much lower than 1 (Weller, 2001).

The LA strategy considers only the LD that exists within a family or cross; this can extend for dozens of cM and is interrupted by recombination after a few generations. This approach uses a limited number of markers per chromosome; therefore, due to recombination between distant markers and the QTL, the association between the markers and the QTL will persist only within a family and for a limited number of generations. This strategy allows QTL mapping of a large interval on the chromosome unless a large number of individuals per family are used. The formula devised by Darvasi and Soller (1997) illustrates this. In a specific genetic map of cattle created by high marker densities, the confidence interval (CI) is given as $CI = 3000/(kns^2)$, where k is the number of informative genitors per individual (1 for families with half siblings and 2 for families with full siblings and F_2 populations), n is the number of genotyped individuals, s is the effect of allelic substitution in connection with the favorable QTL and 3000 cM is the size of the bovine genome (in this species, 1 cM contains approximately eight genes).

Based on this equation and a QTL that segregates with an s equal to a 0.5 residual standard deviation, a family of 1000 half-siblings will have a 95% CI of 12 cM. The following points should be considered for this large CI: (1) if the goal is to use a gene candidate approach within this interval, a large number of genes will need to be sequenced and studied (80 genes, assuming a total of 20,000 genes in a 3000 cM genome) and (2) if the goal is to use MAS, the linkage between the marker and the QTL will not be sufficiently close to guarantee that the association between the marker and the QTL will be consistent across the population. In this latter case, the marker-QTL linkage phase within each family should be established before undergoing MAS (Hayes, 2008). For example, an individual in the population could carry allele M of the marker linked to the favorable allele of the QTL, and another individual from the same population but from a different family could carry allele m for the marker linked to the same favorable QTL allele.

The LA approach is based on associations between marker alleles and the phenotypic QTL classes and, until recently, was widely used because the numbers of markers identified in various species were low and the cost of genotyping was very high. With the recent advent of SNP markers, which exist in large numbers and can be inexpensively genotyped, the use of high-density markers in the genome became possible and finding markers close to the actual QTL became viable. Thus, adoption of the LDA approach, which is superior to LA, became possible.

The LDA strategy is based on linkage disequilibrium between a marker and a QTL in a population and not just within a family, as is done in LA. For this to occur, the marker and QTL must be linked very closely. When this occurs, the association between them is a property of the entire population and will persist for many generations.

Meuwissen and Goddard (2000) showed that the confidence interval could be reduced to 1 cM with LDA mapping. If a QTL polymorphism is due to a recent mutation or to a recent introduction of another population, the LD between a QTL and closely linked markers can be detected at the population level. The closer the marker to the QTL, the greater will be the LD. The confidence interval can be further reduced by combining the LA and LDA strategies and by multiple trait analysis (Meuwissen and Goddard, 2004).

Association analysis is used for fine mapping and is based on population level LD. Linkage can occur when the gene directly affects a trait and when there is LD between the marker and the gene controlling the trait. In the first case, the effect of the gene is measured directly and the marker is functional. The functional mutations are known as quantitative trait nucleotides (QTN). In the second case, the linkage test requires LD between the marker and the QTL. When a mutation occurs on a given chromosome, it creates a haplotype with adjacent loci on the chromosome. In the subsequent generation, this mutation tends to occur within the same haplotype unless there is recombination. This creates the LD used for association mapping.

GENOME-WIDE ASSOCIATION STUDIES (GWAS)

Coefficients and Measurements of Linkage Disequilibrium

As seen previously, LD is defined as the non random association of alleles at different loci. For example, take a locus with alleles A and a and another locus with alleles B and b. The gametic disequilibrium is given by $D = \text{prob}(AB)$ $\text{prob}(ab) - \text{prob}(Ab)\ \text{prob}(aB)$, where prob denotes the probability or frequency of the respective haplotypes. Thus, disequilibrium exists (D is not zero) when the gametes in coupling occur at a different frequency than those in repulsion. Positive values for D show that the gametes in coupling are in excess. Negative D values show that the gametes in repulsion are in excess. After t generations of random crosses, $D_t = D_0(1-r)^t$, and therefore $t = (\log D_t)/[\log D_0(1-r)]$ calculates the number of generations needed to reach equilibrium, where D_0 is the initial disequilibrium and r is the recombination rate.

Take the following allelic frequencies: $p(A) = p_1$; $p(a) = p_2$; $p(B) = q_1$ and $p(b) = q_2$. These have the following equalities: $D = \text{prob}(AB)\ \text{prob}(ab) - \text{prob}(Ab)$ $\text{prob}(aB) = P_{11}\ P_{22} - P_{12}\ P_{21} = p_1q_1\ p_2q_2 - p_1q_2\ p_2q_1 = P_{11} - p_1q_1 = P_{22} - p_2q_2 = p_1q_2 - P_{12} = p_2q_1 - P_{21}$. Thus, the maximum and minimum values for disequilibrium are given by $D\text{max} = \min(Ab, aB) = \min(p_1q_2, p_2q_1)$ and $D\text{min} = \max(AB, ab) = \max(-p_1q_1, -p_2q_2)$.

As an example, consider the following: two loci with two alleles are segregating in the population, and the following information is given: $\text{prob}(AB) = 0.35$, $p(A) = 0.7$, and $p(b) = 0.4$. Is this population in gametic equilibrium? Based on the provided information, we have $p(B) = 1 - 0.4 = 0.6$, and the expected probability of AB is $P(AB) = p(A)\ p(B) = 0.7 \times 0.6 = 0.42$. Thus, $D = \text{prob}(AB) - p(A)$ $p(B) = P_{11} - p_1q_1 = 0.35 - 0.42 = -0.07$. Therefore, the population is in LD and there is an excess of gametes in repulsion. Assuming that the linked loci have a recombination rate of 2%, the number of generations until the disequilibrium falls to one half ($D_t/D_0 = 0.5$) is given as $D_t/D_0 = (1-r)^t = 0.5$. Thus, $0.5 = (1-r)^t$ and $0.5 = (1-0.02)^t$, and solving for t yields $t = 34.31$ generations.

The LD statistic shown above, $D = \text{prob}(AB)\ \text{prob}(ab) - \text{prob}(Ab)\ \text{prob}(aB)$, is very sensitive to individual allele frequencies and therefore cannot be used to compare LD between multiple pairs of loci that cover various locations across the genome. The r^2 statistic developed by Hill and Robertson (1968)[1] is more adequate because it is less dependent on allelic frequencies. This statistic is given by $r^2 = D^2/[\text{prob}(A)\ \text{prob}(a)\ \text{prob}(B)\ \text{prob}(b)]$. Values for r^2 vary from zero (pairs of loci with no disequilibrium between them) to 1 (pairs of loci in complete LD). In the example above, the following haplotype frequencies are observed: $P(AB) = 0.35$; $P(ab) = 0.05$; $P(aB) = 0.25$; $p(Ab) = 0.35$. Therefore, $D = P(AB)\ P(ab) - P(Ab)\ P(aB) = -0.07$ and $D^2 = 0.0049$. The value of r^2 is thus given by $r^2 = D^2/[\text{prob}(A)\ \text{prob}(a)\ \text{prob}(B)\ \text{prob}(b)] = 0.0049/[(0.7)\ (0.3)\ (0.6)$ $(0.4)] = 0.0972$. This level of disequilibrium is considered to be low. Moderate r^2 values are on the order of 0.2 or higher (Hayes et al., 2006).

Another measure for LD is the statistic $D' = $ modulus $(D)/D$max, proposed by Lewontin (1964), which standardizes D by Dmax. The Dmax is given by Dmax $= \min(p_1q_2, p_2q_1)$ if $D > 0$ and Dmax $= \min(p_1q_1, p_2q_2)$ if $D < 0$. This measure of LD is not very accurate because it can be inflated when estimated from small samples or in situations with low allelic frequencies (McRae et al., 2002). Another limitation of D' is its inability to predict the necessary marker density to completely cover the genome by LD.

Thus, the r^2 statistic is preferred. The genetic significance of the r^2 between a marker and an unobserved QTL is that the r^2 measures the proportion of the variation caused by the QTL alleles as explained by the marker alleles. Thus, r^2 decreases as the distance increases, therefore indicating the numbers of markers and phenotypes necessary to make accurate predictions in the context of GWS and QTL detection by LD at the population level. Sample sizes should increase at a rate given by $1/r^2$ to detect an unobserved QTL relative to the sample needed to measure the own QTL (Pritchard and Przeworski, 2001).

The measures of disequilibrium presented herein specifically address loci with two alleles or bi-allelic markers. This is sufficient for SNP markers, but these measures can be extended to multi-allelic markers such as microsatellites. Thus, an estimator for multi-allelic linkage disequilibrium, proposed by Zhao et al. (2005), uses the x^{2*} statistic given by $\chi^{2*} = [1/(m-1)]\sum_{i=1}^{k}\sum_{j=1}^{n}\{D_{ij}^2/[p(a_i)p(b_j)]\}$, where $D_{ij}^2 = p(a_ib_j) - p(a_i)p(b_j)$ and $p(a_i)$ and $p(b_j)$ are the frequencies of the alleles i and j of the markers a and b, respectively. Additionally, $p(a_ib_j)$ represents the frequency of the haplotype (a_ib_j). The quantity m represents the minimum of the number of alleles for the markers a and b. The x^{2*} statistic is a generalization of r^2, and $x^{2*} = r^2$ for bi-allelic markers. Simulations performed by Zhao et al. (2005) show that x^{2*} is the best predictor of the proportion of the variance caused by alleles of the QTL explained by the markers.

The r^2 statistic developed by Hill and Robertson (1968), given by $r^2 = D^2/$ [prob(A) prob(a) prob(B) prob(b)], has an expectation or expected value that is shown by the following equation by Sved (1971): $E(r^2) = 1/(4\ Ne\ L + 1)$. This equation can also be expressed as a function of the recombination rate r in Morgans, resulting in $E(r^2) = 1/(4\ Ne\ r + 1)$. Thus, based on the effective population size (Ne) and the recombination rate, the r^2 can be inferred. Inferred r^2 values are important for calculating the accuracy of GWS.

In exogamous domestic species (animals and perennial plants that are preferentially allogamous), the reduced effective population size is the main cause of LD. In such cases, the expected value for disequilibrium in a given chromosomal segment of size L (in Morgans) can be calculated by the expression $E(r^2) = 1/(4\ Ne\ L + 1)$. The Sved equation shows that the LD decreases rapidly as the distance between the genes increases or as the size of the segment in question increases. This reduction becomes even larger as the effective population size increases (Table 4.1).

For the effective population sizes used in perennial plant improvements (30 to 100), sufficiently large LDs (equal to or greater than 0.2) for QTL selection

TABLE 4.1 Expected Values ($E(r^2)$) for the Linkage Disequilibrium between Two Loci as a Function of the Effective Population Size (Ne) and Length (L) of the Chromosomal Segment That Separates the Two Loci

Ne	L (Morgan)	L (centiMorgan)	$E(r^2)$	Ne	L (Morgan)	L (centiMorgan)	$E(r^2)$
10	0.005	0.5	0.83	100	0.005	0.5	0.33
10	**0.01**	**1**	**0.71**	**100**	**0.01**	**1**	**0.20**
10	0.02	2	0.56	100	0.02	2	0.11
10	0.03	3	0.45	100	0.03	3	0.08
10	0.04	4	0.38	100	0.04	4	0.06
10	0.05	5	0.33	100	0.05	5	0.05
20	0.005	0.5	0.71	200	0.005	0.5	0.20
20	**0.01**	**1**	**0.56**	**200**	**0.01**	**1**	**0.11**
20	0.02	2	0.38	200	0.02	2	0.06
20	0.03	3	0.29	200	0.03	3	0.04
20	0.04	4	0.24	200	0.04	4	0.03
20	0.05	5	0.20	200	0.05	5	0.02
30	0.005	0.5	0.63	500	0.005	0.5	0.09
30	**0.01**	**1**	**0.45**	**500**	**0.01**	**1**	**0.05**
30	0.02	2	0.29	500	0.02	2	0.02
30	0.03	3	0.22	500	0.03	3	0.02
30	0.04	4	0.17	500	0.04	4	0.01
30	0.05	5	0.14	500	0.05	5	0.01
50	0.005	0.5	0.50	1000	0.005	0.5	0.05
50	**0.01**	**1**	**0.33**	**1000**	**0.01**	**1**	**0.02**
50	0.02	2	0.20	1000	0.02	2	0.01
50	0.03	3	0.14	1000	0.03	3	0.01
50	0.04	4	0.11	1000	0.04	4	0.01
50	0.05	5	0.09	1000	0.05	5	0.00

can be obtained with markers that are spaced 1 to 3 cM apart. The r^2_{mq} or $E(r^2)$ is a weighted average of the r^2 of each marker-QTL pair, being r^2 the square of the correlation (r) between the alleles or genotypes present at the marker locus and the QTL locus (Table 4.2).

TABLE 4.2 Calculations of the Linkage Disequilibrium between a Marker and QTL

Individual	Num. alleles at marker locus (X_a)	Num. alleles at QTL locus (X_b)
1	0	0
2	2	1
3	1	1
4	1	0
5	2	1
Correlation (r)	$r = 0.76$	$r^2 = 0.58$

The correlation coefficient between two variables or alleles at loci a and b is given by

$$r = \frac{Cov(a,b)}{[Var(a)Var(b)]^{1/2}} = \frac{\sum ab - \sum a \sum b}{[Var(a)]^{1/2}[Var(b)]^{1/2}}$$
$$= \frac{Pr\,ob(ab) - Pr\,ob(a)\,Pr\,ob(b)}{[pq]^{1/2}[rs]^{1/2}} = \frac{D}{[pq\,rs]^{1/2}}$$

The square of this quantity equals $r^2 = \frac{D^2}{[pq\,rs]}$, which is the standard measurement for LD. Using the incidence matrix X for the markers, the value of r can be expressed as

$$r_{(a,b)} = \frac{Cov(X_{ia}, X_{ib})}{[Var(X_{ia})]^{1/2}[Var(X_{ib})]^{1/2}}$$

D is defined by $D = \text{Prob}(ab) - \text{Prob}(a)\,\text{Prob}(b)$, where $\text{Prob}(a)$ is the frequency of allele a and $\text{Prob}(ab)$ is the frequency of genotype ab. Generically, p, q, r, and s are the frequencies of the alleles A, a, B, and b, respectively. The equation $Var(a) = pq$ assumes a Bernoulli distribution for the presence of an allele.

The relationship between the genetic effects of a marker and a QTL can be better understood with the following models: genetic effect of the QTL on the phenotype (g_{QTL}), $y = u + g_{QTL} + e$ and genetic effect of the marker on the phenotype (g_m), $y = u + g_{QTL} + e = u + Xg_m + e$. The variable g_m is a regression coefficient given by

$$g_m = Cov(y, X)/Var(X) = Cov(g_{QTL}, X)/Var(X)$$
$$= r[Var(g_{QTL})/Var(X)]^{1/2} = r\{Var(g_{QTL})/[2p(1-p)]\}^{1/2}$$

The amount of variation in the QTL explained by the marker is given by

$$Var(Xg_m) = g_m^2 Var(X) = r^2[Var(g_{QTL})/Var(X)]Var(X) = r^2 Var(g_{QTL})$$

Thus, we find that r^2 is the proportion of variance in the QTL explained by the marker.

The extent of the LD depends on recent and historical recombinations as well as the current and past Ne. Domesticated plant and animal populations have lower current Ne than past Ne. In humans, the contrary is true due to the current rapid increase in the population. Hayes et al. (2003) stated that LD in short chromosomal segments (short distances) depends on the historical effective population size many generations ago. However, disequilibrium across long distances depends on the recent history of the population. As linear changes in populations occur, it must follow that the measure r^2 reflects the Ne associated to $1/(2L)$ generations back. Therefore, the expected r^2 when Ne changes over time is given by $E(r^2) = 1/(4 \, Ne_t \, L + 1)$, where $t = 1/(2L)$. In humans, the Ne equals approximately 10,000 (Kruglyak, 1999). In domesticated animals and perennial plants, the Ne can be lower (in the order of 100). Therefore, the LD should be lower in humans. However, in the past, the Ne of the human population was low. Thus, for long distances between markers, the r^2 values are lower in humans than in domesticated plant and animal species. And the r^2 values for short distances between markers are more similar in humans and domesticated animal species. Moderate LDs (r^2 greater than or equal to 0.2) in humans extend for fewer than 5 kb or 0.005 cM. In cattle, moderate LDs extend up to 100 kb. However, high LD values (r^2 greater than or equal to 0.8) extend for only very short distances in both humans and cattle (Tenesa et al., 2007).

Dutch and Australian dairy cattle populations show a similar decline in LD because these populations are related by origin and have similar histories and Ne values. Norwegian red cattle (Ne equal to 400) have a faster decline in LD than Dutch dairy cattle (global Ne of 150). The different Ne values justify the different LDs in both populations (Zenger et al., 2007).

Methods for QTL Analysis via LDA

For a long time, mapping studies were based on linkage analyses associated with pedigree data. More recently, methods based on LD in unrelated individuals have been recommended as powerful tools with which to produce precise estimates of gene locations. These methods are based on the premises listed below. When a novel allele is introduced into the population either by mutation or migration, it exists in the population with a group of marker alleles. The length of this haplotype is reduced over several generations due to recombination events and, after many generations, only the markers in the immediate neighborhood of the new allele locus are likely to remain on the same chromosome segment.

Due to the low LD present in these populations, the associations of markers with the trait are harder to find and extremely high marker density is required to increase the probability that a marker will be in LD with the causal loci. Despite this disadvantage, whenever markers are identified to be associated with an allele that influences a given trait, a strong correlation between the trait and the marker should indicate that the coding locus for the trait is located very near the marker or the identified marker could in fact be QTN.

LDA mapping seeks to increase the precision of QTL position estimates because, in some situations, the meiosis number in the genotyped pedigree is not sufficient for the LA to be precise. LDA methods provide fine mapping based on quantified LD in the gametic phase which is persistent across families in an allogamous population. In this case, the linkage phase does not change between families or between generations. This method is based on the fact that in a small population, the founders have a small number of distinct haplotypes and, at highly linked loci, there is not sufficient time for recombination to break the associations between markers and the mutations that affect the QTL (Perez-Enciso et al., 2003). This type of mapping is also known as association mapping, which became possible with the advent of high-density SNP and DArT markers. The strategy of population-level trait-marker association depends on small blocks of genes in disequilibrium, and therefore the resolution is very high (small distances between genes). Although the resolution is higher, QTL detection and precision mapping require a large number of markers. Association mapping uses the general population rather than a specific mapping population. The association between a marker and a QTL depends on the recombination frequency between them. To find a marker reasonably close to a QTL, there must be a low recombination rate. The larger the LD, the closer the marker is to the gene, and this LD or association will be valid even for genetically distant individuals.

Two approaches, genomic scanning and candidate genes, can be used in association genetics or mapping. In the latter approach, only markers in the individual candidate genes are used. For association genetics, the mapping population should be large and have a large degree of LD. LDA mapping uses genomic scanning with high-density markers (1 marker per 0.5 to 2 cM). The success of the method depends on the amount of LD in the population. Because markers can be in incomplete LD with the QTL, both the association between markers and QTLs in the population and the co segregation of markers and QTL within families can be used simultaneously for QTL detection via the LDA-LA method, which combines the properties of LD (in linkage disequilibrium) and LE (in linkage equilibrium) markers, respectively.

LDA-based mapping is performed by calculating the probabilities that the haplotypes shared by individuals are identical by descent from a common ancestor, conditional on marker data. Accurate determination of the linkage phases and QTL genotypes is necessary for fine mapping. Thus, a pure LDA analysis might result in a high number of false positives or false inferences of association in the absence of linkage. Therefore, methods (LA-LDA) that simultaneously incorporate information about population LD and linkages within families are recommended to mitigate the effects of spurious associations between the markers and QTLs (Meuwissen and Goddard, 2004).

MAS and GWS are increasingly effective as the markers become closer to the QTLs. Given the small distances between the genes on chromosomes, the accurate mapping of QTLs is difficult. On average, a 10-cM chromosomal segment can contain approximately 200 genes. Thus, a high density of genotyped markers increases the QTL mapping resolution. However, if the aim is to find the actual gene that affects a trait, the confidence interval for the QTL remains large, even for a QTL with a large effect and when using a large sample size (Weller, 2001). LDA mapping strategies are described below.

GENOME-WIDE MAPPING VIA SINGLE MARKER REGRESSION

To identify statistically significant effects, genome-wide association studies (GWAS) seek out associations between loci and phenotypic traits in a population by hypothesis testing. The following regression model for single markers can be used to find QTL-marker associations in a panmictic population (Resende, 2008): $y = l\mu + Xm_i + e$, where y is the vector of observed phenotypes, l is a vector with the value of 1, μ is a scalar with the overall mean, m_i is the fixed effect of one of the bi-allelic marker alleles, and e is the vector of random residuals. X is the incidence matrix for m_i. This model assumes that the marker will affect the trait only if it is in LD with the putative QTL. Other fixed and random effects can be incorporated into this model. For example, consider the analysis of 12 individuals for a particular trait and an SNP marker. The genotypic and phenotypic data for the individuals are shown in Table 4.3.

The incidence matrix X connects the number of each SNP allele to the individual phenotypes. Only the effects of one of the alleles need to be included. Thus the matrix X will have only one column for the effects of one of the SNP alleles, such as A. This column contains the number of copies of A possessed by the individuals. Therefore, it has values of 0, 1, or 2 for a diploid individual. The number of rows in the matrix is equal to the number of individuals.

The matrix 1 includes a column for the overall mean. The matrices 1 and X (number of A alleles), shown as transposed matrices, are $1'_{(12x1)} = [1\,1\,1\,1\,1\,1\,1\,1\,1\,1\,1\,1]$ and $X'_{(12x1)} = [1\,2\,1\,2\,1\,0\,2\,1\,0\,2\,1\,2]$. The least squares equations for estimating the effects of the overall mean and the SNP is:

$$\begin{bmatrix} 1'1 & 1'X \\ X'1 & X'X \end{bmatrix} \begin{bmatrix} \hat{u} \\ \hat{m}_i \end{bmatrix} = \begin{bmatrix} 1'y \\ X'y \end{bmatrix},$$

where y is the phenotype vector. Solving this system yields:

$$\begin{bmatrix} \hat{u} \\ \hat{m}_i \end{bmatrix} = \begin{bmatrix} 7.2713 \\ 3.7856 \end{bmatrix}$$

TABLE 4.3 Genotypic and Phenotypic Data for 12 Individuals for a Particular Trait and an SNP Marker

Individual	Phenotype	First SNP1 allele	Second SNP1 allele
1	9.87	A	a
2	14.48	A	A
3	8.91	A	a
4	14.64	A	A
5	9.55	A	a
6	7.96	a	a
7	16.07	A	A
8	14.01	A	a
9	7.96	a	a
10	21.17	A	A
11	10.19	A	a
12	9.23	A	A

The null hypothesis, which states that the marker does not have any effect on the trait, can be evaluated with the F-test. The null hypothesis is rejected if $F > F(a, v_1, v_2)$, where F is the *Snedecor* statistic calculated from the data, a is the significance level and v_1 and v_2 are the degrees of freedom for the F distribution. The alternative hypothesis is that the marker affects the trait because the marker and the QTL are in LD. The F-statistic value is calculated as

$$F = \frac{QM \operatorname{Re} gression}{\hat{\sigma}_e^2} = \frac{\hat{m}X'y + \hat{u}1'y - (1/n)(1'y)^2}{(y'y - \hat{m}X'y - \hat{u}1'y)/(n-2)}$$

In this example, the calculated value for F was 9.74. This value can be compared to the F-value for a 5% significance level and 1 and 10 degrees of freedom on an F-table, which gives 4.96. Thus, the SNP effect is significant. This was expected because the individuals with higher phenotypic values were those with a larger number of A alleles for the SNP, as clearly shown in the data table.

STATISTICAL POWER AND SIGNIFICANCE OF ASSOCIATION FOR QTL DETECTION

The power of the marker-QTL linkage test depends on the following factors (Pritchard and Przeworski, 2001; Meuwissen et al., 2002; Hayes et al., 2006; Fernando et al., 2004; Macleod et al., 2010):

1. The r^2 (statistical measurement of the LD) between the marker and QTL. The genetic significance of the r^2 between a marker and an unobserved QTL is that it measures the proportion of the variance caused by alleles of the QTL explained by the marker alleles. The sample size required to detect an unobserved QTL should increase at a rate of $1/r^2$, compared to the sample needed to analyze the QTL itself.
2. The proportion of phenotypic variance (σ_f^2) explained by the QTL; in other words, the coefficient of determination for the q QTL effect ($h_q^2 = \sigma_q^2/\sigma_f^2$).
3. The number of individuals analyzed.
4. The chosen significance level.
5. The frequency, p, of the rare marker allele, which determines the minimum number of observations necessary to estimate the allelic effect. When p is less than 0.1, the power is sensitive to this allelic frequency.

The power of a test is the probability of rejecting H_0 when H_0 is false, or the ability to detect a QTL in the population when it exists. The power of a test to detect a QTL at various r^2 values for a QTL-marker pair can be calculated with a formula derived by Luo (1998). To achieve a power greater than or equal to 80% for detecting a QTL with an h_q^2 equal to 0.05 based on 1000 phenotypic observations, an r^2 of at least 0.2 is required. This result assumed that the frequency of the rare allele is higher than 0.2.

Macleod et al. (2010) reported that the power of QTL detection with an h_q^2 of 5% and 365 genotyped individuals was 37% ($p<0.001$). The authors also found a strong correlation between the F-values of significant SNPs and their r^2 with the QTL. The correlation between the Snedecor F-statistic and D' was practically zero.

When making a conclusion, a researcher commits a type I error if he/she rejects a true H_0 hypothesis and a type II error if he/she accepts a false H_0 hypothesis. The probability of committing a type I error is called α, and the largest value of α for a true H_0 is called the significance level of a statistical test; in other words, the significance of a test is the maximum probability allowed for making a type I error.

Choosing the significance level to be used in GWAS requires several considerations. This is because thousands of markers will be tested and therefore one confronts the problem of determining the nominal significance levels to be used with multiple tests. Thus, the nominal significance level for each test will not correspond to the level used for the experiment as a whole. At a significance level of 5%, it is expected that 5% of the results will be false positives. If 20,000 markers are tested, 1000 false positives are expected. The Bonferroni correction can mitigate this problem. However, this correction does not account for the fact that tests on the same chromosome are not independent because the markers can be in LD between themselves and also with the QTL.

The permutation test technique was proposed by Churchill and Doerge (1994) to address the challenge of multiple tests in QTL mapping experiments. This technique is used to establish appropriate significance levels. Hoggart et al.

(2008) derived an explicit approximation for the type I error rate, thus avoiding the need for permutation procedures. Another option to avoid false positives is to track this number in comparison with the number of positive results, as done in Fernando et al. (2004). The researcher can thus establish a significance level based on an acceptable proportion of false positives.

In addition to the correction techniques listed above, there is the false discovery rate (FDR), which is defined as the expected proportion of detected false positive QTLs. The FDR can be calculated as FDR $= m$ Pmax/n, where Pmax is the largest QTL P-value that exceeds the significance level, n is the number of QTLs that exceed the significance threshold, and m is the number of tested markers (Weller, 2001). If 10,000 SNPs are tested with a significance level (P-value) of 0.001 and 80 SNPs are considered significant, then FDR $= 10,000 \times$ 0.001/80 $= 0.125$. This result (12.5%) is considered acceptable for the FDR.

An alternative approach to decreasing the false positive rate is the use of a model that includes a vector of polygenic effects, which includes a relationship matrix and permits correction for the population structure. Macleod et al. (2010) reported an increased number of false positives (type I errors) when polygenic effects were not included in the model. In this case, it is recommended that markers be used to infer the relationship matrix as shown by Hayes et al. (2007). For a given marker locus, the genetic similarity, Sxy, between two individuals x and y is calculated as follows (Hayes et al., 2006, 2007):

1. $Sxy = 1$ when the genotype $x = ii$ (both alleles at the locus are identical) and the genotype $y = ii$ or when $x = ij$ and $y = ij$;
2. $Sxy = 0.5$ when the genotype of $x = ii$ and the genotype of $y = ij$ or vice versa;
3. $Sxy = 0.25$ when the genotype of $x = ij$ and the genotype of $y = ik$;
4. $Sxy = 0$ when the two individuals do not share alleles at the locus.

Similarity arising by chance is represented by

$$Sa = \sum_{i=1}^{g} p_i^2,$$

where p is the allele frequency in the population and g is the number of alleles at the locus. The relatedness between individuals x and y for the locus is thus expressed as $r = (Sxy - Sa)/(1 - Sa)$. The average relatedness between the individuals is calculated as the average r across all of the loci. Thus, when a large number of markers are present, the resulting marker relationship matrix can capture the effects of Mendelian segregation, which are not included in pedigree-based relationship matrices.

Cross-validation methods can be used to estimate confidence intervals in GWAS. For these methods, the data are divided into two halves, and the association analysis is performed three times, once for each half of the data and once with all of the data. The 95% confidence interval for the QTL position is given by the position of the most significant SNP in an analysis of the complete dataset $\pm 1.96s$, where s is the standard error of the QTL and is calculated from the equation

$$s = \left(\frac{1}{4n} \sum_{i=1}^{n} x_{1i} - x_{2i} \right)^{1/2}$$

for n SNP pairs with significant effects. The components x_{1i} and x_{2i} are the positions of the most significant SNP in each of the half-datasets for the ith most significant QTL position in the complete dataset. This is valid when the analysis of each half-dataset confirms an SNP that was declared to be significant in the complete dataset analysis (Hayes et al., 2006, 2007).

GENOME-WIDE MAPPING WITH HAPLOTYPE MIXED MODELS

Haplotypes are specific combinations of multiple linked markers and can be considered as alleles of a "super-locus." Haplotypes can be used in place of simple markers in GWAS. An advantage is that they can exist in greater disequilibrium with the QTLs. When this occurs, the r^2 is larger and therefore the power of the experiment is increased. The proportion of the QTL variance explained by the markers can be calculated as follows (Hayes et al., 2006): If q_1 and q_2 are the frequencies of the two QTL alleles, the markers can be grouped into n haplotypes with the frequency p_i for the ith haplotype. This can be shown in a contingency table (see Table 4.4).

For the haplotype i in the data, the LD is calculated by $D_i = p_i(q_1) - p_i q_1$, where $p_i(q_1)$ is the proportion of i haplotypes in the dataset that have allele 1 of the QTL (observed in the data), p_i is the proportion of i haplotypes, and q_1 is the frequency of QTL allele 1. The proportion of the variance in the QTL explained by the haplotypes, when corrected for sampling effects, can be calculated by

$$r^2(h, q) = \frac{1}{q_1 q_2} \sum_{i=1}^{n} D_i^2 / p_i$$

TABLE 4.4 Contingency Table

	Haplotypes			
	1	i	n	Total
QTL allele 1	$p_1 q_1 - D_1$	$p_i q_1 - D_i$	$p_n q_1 - D_n$	Q_1
QTL allele 2	$p_1 q_2 + D_1$	$p_i q_2 + D_i$	$p_n q_2 + D_n$	Q_2
Total	p_1	p_i	p_n	1

Thus, the r^2 depends on the LD, the frequency of haplotype h, and the frequencies of the QTL alleles. The r^2 values can be obtained by simulating the different frequencies q_1 and q_2, the genome size, and the haplotypes. The larger the effective population size, the smaller the proportion of genetic variance that will be explained by the haplotypes.

The following mixed linear model is used to estimate the effects of the haplotypes: $y = 1'u + Xh + Za^* + e$, where y is the observed phenotypes vector, u is the scalar of the average (fixed effect), h is the random effects haplotypes vector (intervals), a^* is the polygenic effects vector (random), and e is the random residuals vector. X and Z are the incidence matrices for h and a^*. The effects of the haplotypes should preferably be treated as random because they occur in large numbers and some occur a limited number of times (haplotypes with a small number of observations should be penalized by the shrinkage effect).

The magnitude of h is equal to the number of intervals multiplied by 4 (number of possible haplotypes for each interval). The incidence matrix X contains the values 0, 1, and 2, which represent the number of alleles (of the putative QTL) or type h_i haplotypes in a diploid individual. The algebraic details of this model were previously explained by Resende (2008). The additive variances of the genes ($\sigma_{a^*}^2$) and the haplotypes (σ_h^2) can be estimated by the restricted maximum likelihood applied on phenotypic data or by the variance between haplotypes or the variance of the chromosomal segments of the QTL. The significance of the haplotypic effects is evaluated with the likelihood ratio test. For mapping, the fit of the described model relies on estimations of the variance σ_h^2 and LRT to test their significance. There are no specific interests for the best linear unbiased prediction effect for h, which are emphasized and utilized in the MAS.

GWAS IN HUMANS

The first studies of human quantitative genetics that sought to understand genetic control of traits were based on estimates of heritability (h^2) by analyzing pairs of twins according to the idea of pedigree-based similarity between relatives (pedigrees: alleles that are identical by descent (IBD)). This approach considers all of the loci or genes, both the common and rare variants (low-frequency genes), that control the trait or the total h^2. The role of individual genes in genetic trait control was later studied according to the Fulker and Cardon (1994) method and by estimating the h^2 of a marker locus in the context of QTL mapping, as described by Resende (2008) and Cruz et al. (2009). This method is based on LA within full sib families and uses two molecular markers at a time.

Visscher et al. (2006) proposed an approach to estimating h^2 by simultaneously using all of the marker loci and also using segregation analysis within full sib families. This genome-wide approach is also based on IBD and takes advantage of the exact relationships. In humans, the estimated h^2 was 0.80 for height. This method includes both common and rare variants (all of the genes or the total h^2) because it also incorporates the pedigree by genotyping the parents and estimating IBD alleles for all of the loci. Another method for studying trait control at a population level, not only within families, is GWAS. This approach is based on linkage disequilibrium in the population, but only uses one marker locus at a time via a fixed regression analysis of unrelated individuals. It thus seeks to identify significant markers that can be used in the estimation. In humans, the h^2 captured by the significant markers was only 0.10 for height.

When applied to a family with full siblings, GWAS can be described as linkage analysis. In this analysis, markers at some distance from a QTL will exhibit an association with the traits because only one generation of recombination occurred between the progenitors and the siblings. Consequently, a marker allele and a QTL allele on the same chromosome will tend to be inherited together. A more effective procedure (GWAS-SE) for capturing the majority of the heritability of a trait is a population LDA in which all markers are analyzed simultaneously (SE), a method similar to GWS. This method is based on random regression for the prediction of latent QTL effects. It uses unrelated individuals, although all of the individuals of a species are related to some degree because they share common ancestors and thus identical alleles in state (IBS), not necessarily declared as IBD due to a limitation of having recorded a really complete pedigree.

SNP markers capture this ancestral relatedness and therefore estimate genetic relationships by IBS (Powell et al., 2010; Visscher et al., 2010). Population genetics (LA, LDA, and genetic mapping) and quantitative genetics (estimation of heritability) have traditionally been used separately in human genetics. By combining these two areas, GWS allowed an h^2 measurement of 0.45 for height in humans. The remaining portion $(0.80 - 0.45 = 0.35)$ was not captured due to many low-frequency variants (including loci with large effects). The genetic variance at a locus i is given by $\sigma_{ai}^2 = 2p_i(1 - p_i)a_i^2$, ignoring dominance. Thus, a rare allele cannot explain a large part of the genetic variance, even if it has a large effect. A large sample size is needed for these loci to be captured by markers and detected. In GWS experiments, the total additive genetic variance is estimated by $\sigma_a^2 = \sum_i 2p_i(1 - p_i)a_i^2$.

Aulchenko et al. (2007) proposed the GRAMMAR method for multiple-stage GWAS, as described below. After fitting the model $y = Xb + Zg + e$, one obtains $\hat{e} = y - X\hat{b} - Z\hat{g}$, where g is a polygenic effect vector. The model $\hat{e} = 1u + Wm_i + e$ can then be fit to identify significant markers. The model $y = Xb + Wm_i + Zg + e$ is then adjusted by using only the significant SNPs. This reduces the necessary computation time when calculating thousands of markers. The effects of m are fit as the fixed effects because the SNPs do not model the familiar structure in g; in other words, they do not explain the correlation between related individuals by IBD. The method is based on the fact that the effects of the major genes are included in the conditional residual vector after fitting g under an infinitesimal polygenic model (fitting or elimination of family effects or variations between the pedigree or population structure). The final analysis returns to the complete model. This time, the polygenic effect is included to correct the data for the family structure with the relationship matrix A, as $g \sim N(0, A\sigma_g^2)$.

CAPTURING h^2 IN HUMANS WITH IMPERFECT LD BETWEEN SNPS AND CAUSAL VARIANTS

Visscher et al. (2010) addressed the GWAS results for height in humans. The h^2 captured by GWAS in traditional studies was approximately 0.10. This low value was obtained because low frequency variants (MAF < 0.10) were not in perfect LD with common markers (MAF > 0.10); in such instances, the r^2 is low and variants with small effects are not detected significantly by traditional GWAS, even when they are in LD with common markers. In a study by Yang et al. (2011), the captured h^2 was 0.45. This occurred because variants with small effects were still not detected significantly, but they were captured by GWS when in LD with common markers because GWS does not use significance for marker effects. The maximum possible value for r^2 is largely determined by the allelic frequencies of the two loci. The more the allelic frequencies are different, the lower the r^2 value. Thus, as most genotype SNPs are common, r^2 will be low if the variants are rare and therefore the variance (σ_{mi}^2) associated with the SNP will be substantially less than the variance (σ_{ai}^2) of the QTL (Visscher et al., 2010). The expressions $r^2 = \sigma_{mi}^2/\sigma_{ai}^2$ and $\sigma_{mi}^2 = r^2\sigma_{ai}^2$ illustrate this point.

In practice, only the LD between the SNPs can be estimated. This estimate is only useful when the SNP and gene have similar allelic frequencies. A gene can be in LD with several SNPs that can collectively capture the causal variant, even if none of the SNPs are in perfect LD with the gene (Visscher et al., 2010). Thus, an SNP can fail to be significant but, together with other SNPs, can still be important when explaining genetic variance and maximizing the selective accuracy. Therefore, it is not recommended that significance tests be applied prior to GWS. Even when using tens of thousands of markers, the markers will not capture all of the genetic variance if the variants are rare and the markers common. Thus, the efficiency of GWS depends on the genetic architecture of the trait in the population. If the trait is governed by a large number of rare variants that explain a large portion of the genetic variance, GWS will be less successful. In such cases, it is recommended that residual polygenic effects be fitted in the model to capture rare variants.

In summary, the following causes for missing heritability are: (1) low-frequency variants (MAF < 0.10) that are not in high LD with common markers (MAF > 0.10), resulting in a low r^2; (2) a small number of markers that cause a low r^2; and (3) the use of only significant SNPs in GWAS. Simultaneous estimation is necessary because the SNPs are in LD and are thus dependent and correlated. Simultaneous regression is equivalent to phenotypic regression in all of the principal components derived from the markers because the amount of the experienced shrinkage for each estimated effect is proportional to its quadratic singular value (Campos et al., 2010). This supports the use of GWAS with the simultaneous estimation (GWAS-SE) method, according to Yang et al. (2011).

GWAS via BayesCπ and BayesDπ

The BayesCπ and BayesDπ methods (described by Habier et al., 2011 and Resende et al., 2011) are advantageous because they provide information on the genetic architecture of the quantitative trait and identify the QTL positions by modeling the frequencies of single nucleotide polymorphisms (SNP) with nonzero effects. They are advantageous over the regression analysis of single markers because they simultaneously account for all markers.

However, care needs to be taken whenever the number of markers is larger than the number of individuals genotyped and phenotyped. Gianola (2013) showed that in these cases, in the Bayesian approaches proposed, such as BayesC and BayesD, the prior is always influential, which could affect the inference to whether or not a marker is associated with the trait.

In the BayesC method, a common variance is specified for all of the loci. The BayesD method maintains specific variances for each locus. Additionally, π is treated as an unknown with a uniform *a priori* distribution $(0,1)$, thus producing the methods BayesCπ and BayesDπ. Modeling π is very interesting in association analysis. The majority of the markers are not in LD with the genes. Thus, a group of markers associated to a trait must be identified. Differently, the BayesB method subjectively determines π. Using the indicator variable δ_i, the BayesCπ and BayesDπ methods model the additive genetic effect of individual j as

$$a_j = \sum_{i=1}^{n} \beta_i x_{ij} \delta_i,$$

where $\delta_i = (0,1)$. The distribution of $\delta = (\delta_1 ... \delta_n)$ is binomial with a probability π. This mixture model is more parsimonious than the BayesB method. According to the model hierarchy, a distribution must be postulated for π and must be a beta distribution, which when appropriately specified becomes a uniform distribution $(0,1)$ (Legarra et al., 2011).

The quantities for x_{ij} are elements of the codominant marker genotype vector and are generally coded as 0, 1, or 2, depending on the number of copies of one of the alleles at marker locus i, and β_i is defined as an element of the vector of the regression coefficients, which includes the marker effects on a phenotypic trait y by means of the LD with the genes that control the trait.

REFERENCES

Anderson, L., Georges, M., 2004. Domestic animal genomes: deciphering the genetics of complex traits. Nature Reviews Genetics 5 (3), 202–212.

Aulchenko, Y.S., Konning, D., Haley, C., 2007. Grammar: a fast and simple method for genome-wide pedigree-based quantitative trait loci association analysis. Genetics, Austin 177, 577–585.

Campos, G.de los, Gianola, D., Allison, D.B., 2010. Predicting genetic predisposition in humans: the promise of whole-genome markers. Nature Reviews Genetics, London 11, 880–886.

Churchill, G.A., Doerge, R.W., 1994. Empirical threshold values for quantitative trait mapping. Genetics 138, 963–971.

Cruz, C.D., Good God, P.I.V., Bhering, L.L., 2009. Mapeamento de QTLs em populações exogâmicas. In: Borém, A., Caixeta, E.T. (Orgs.), Marcadores Moleculares, vol. 1, second ed. Folha de Viçosa, Viçosa, MG, pp. 443–481.

Darvasi, A., Soller, M., 1997. A simple method to calculate resolving power and confidence interval of QTL map location. Behavior Genetics 27, 125–132.

Ewing, B., Green, P., 2000. Analysis of expressed sequence tags indicates 35,000 human genes. Nature Genetics 25, 232–234.

Fernando, R.L., Nettleton, D., Southey, B.R., Dekkers, J.C.M., Rothschild, M.F., Soller, M., 2004. Controlling the proportion of false positives in multiple dependent tests. Genetics 166, 611–619.

Fulker, D.F., Cardon, L.R., 1994. A sib-pair approach to interval mapping of quantitative trait loci. American Journal of Human Genetics 54, 1092–1103.

Gianola, D., 2013. Priors in whole-genome regression: the Bayesian alphabet returns. Genetics: Early Online, published on May 1, 2013 as 10.1534/genetics.113.151753.

Habier, D., Fernando, R.L., Kizilkaya, K., Garrick, D.J., 2011. Extension of the Bayesian alphabet for genomic selection. BMC Bioinformatics 12, 186.

Hartl, L.D., Jones, E.W., 2002. Essential Genetics: a Genomics Perspective. Jones & Bartlet, Sudbury.

Hayes, B.J., 2008. Course on QTL Mapping, MAS and Genomic Selection. Iowa State University, Ames.

Hayes, B.J., Chamberlain, A.J., Goddard, M.E., 2006. Use of markers in linkage disequilibrium with QTL in breeding programs. In: World Congress of Genetics Applied to Livestock Production, 8, 2006. Proceedings. Belo Horizonte: Ed. da UFMG. 1 CD-ROM.

Hayes, B.J., Chamberlain, A.J., McPartlan, H., Macleod, I., Sethuraman, L., Goddard, M.E., 2007. Accuracy of marker assisted selection with single markers and marker haplotypes in cattle. Genetical Research 89, 215–220.

Hayes, B.J., Visscher, P.E., McPartlan, H., Goddard, M.E., 2003. A novel multi-locus measure of linkage disequilibrium and it use to estimate past effective population size. Genome Research 13, 635–643.

Hill, W.G., Robertson, A., 1968. Linkage disequilibrium in finite populations. Theoretical and Applied Genetics 38, 226–231.

Hoggart, C.J., Whittaker, J.C., De Iorio, M., Balding, D.J., 2008. Simultaneous analysis of all SNPs in genome-wide and re-sequencing association studies. PLoS Genetics 4 (7), e1000130.

Kruglyak, L., 1999. Prospect for whole genome linkage disequilibrium mapping of common disease genes. Nature Genetics 22, 139–144.

Lande, R., Thompson, R., 1990. Efficiency of marker-assisted selection in the improvement of quantitative traits. Genetics 124, 743–756.

Legarra, A., Robert-Granié, C., Croiseau, P., Guillaume, F., Fritz, S., 2011. Improved Lasso for genomic selection. Genetics Research 93 (1), 77–87.

Lewontin, R.C., 1964. The interaction of selection and linkage. II. Optimal models. Genetics 50, 757–782.

Luo, Z.W., 1998. Linkage disequilibrium in a two-locus model. Heredity 80, 198–208.

Macleod, I.M., Hayes, B.J., Savin, K., Chamberlain, A.J., McPartlan, H., Goddard, M.E., 2010. Power of a genome scan to detect and locate quantitative trait loci in cattle using dense single nucleotide polymorphisms. Journal of Animal Breeding and Genetics 127 (2), 133–142. <http://www.ncbi.nlm.nih.gov/pubmed/20433522>.

McRae, A.F., McEvan, J.C., Dodds, K.G., Wilson, T., Crawford, A.M., Slate, J., 2002. Linkage disequilibrium in domestic sheep. Genetics 160, 1113–1122.

Meuwissen, T.H.E., Goddard, M.E., 2000. Fine mapping of quantitative trait loci using linkage disequilibria with closely linked marker loci. Genetics 155, 421–430.

Meuwissen, T.H.E., Goddard, M.E., 2004. Mapping multiple QTL using linkage disequilibrium and linkage analysis information and multitrait data. Genetics Selection Evolution 36, 261–279.

Meuwissen, T.H.E., Hayes, B.J., Goddard, M.E., 2001. Prediction of total genetic value using genome-wide dense marker maps. Genetics 157, 1819–1829.

Meuwissen, T.H.E., Karlsen, A., Lien, S., Olsaker, I., Goddard, M.E., 2002. Fine mapping of a quantitative trait locus for twinning rate using combined linkage and linkage disequilibrium mapping. Genetics 161, 373–379.

Perez-Enciso, M., Toro, M.A., Tenenhaus, M., Gianola, D., 2003. Combining gene expression and molecular marker information for mapping complex trait genes: a simulation study. Genetics 164, 1597–1606.

Powell, J.E., Visscher, P.M., Goddard, M.E., 2010. Reconciling the analysis of IBD and IBS in complex trait studies. Nature Reviews Genetics 11, 800–805.

Pritchard, J.K., Przeworski, M., 2001. Linkage disequilibrium in humans: models and data. American Journal of Human Genetics 69, 1–14.

Resende, M.D.V., 2008. Genômica Quantitativa e Seleção no Melhoramento de Plantas Perenes e Animais. Embrapa Florestas, Colombo. 330p.

Resende, M.D.V., Silva, F.F., Viana, J.M.S., Peternelli, L.A., Resende Junior, M.F.R., et al., 2011. Métodos Estatísticos na Seleção Genômica Ampla. Embrapa Florestas, Colombo. 106p.

Sved, J.A., 1971. Linkage disequilibrium and homozygosity of chromosome segments in finite populations. Theoretical Population Biology 2, 125–141.

Tenesa, A., Navarro, T., Hayes, B.J., Duffy, D.L., Clarke, G.M., Goddard, M.E., et al., 2007. Recent human effective population size estimated from linkage disequilibrium. Genome Research 17, 520–526.

Visscher, P.M., Medland, S.E., Ferreira, M.A.R., Morley, K.I., Zhu, G., et al., 2006. Assumption-free estimation of heritability from genome-wide identity-by-descent sharing between full siblings. PLoS Genetics 2 (3), e41.

Visscher, P.M., Yang, J., Goddard, M.E., 2010. A commentary on Common SNPs explain a large proportion of the heritability for human height by Yang et al. (2010). Twin Research and Human Genetics 13 (6), 517–524.

Weller, J.I., 2001. Quantitative Trait Loci Analysis in Animals. CABI Publishing, London. 287p.

Yang, J., Lee, S.H., Goddard, M.E., Visscher, P.M., 2011. Gcta: a tool for genome-wide complex trait analysis. American Journal of Human Genetics 88, 76–82.

Zenger, K.R., Khatkar, M.S., Cavanagh, J.A., Hawken, R.J., Raadsma, H.W., 2007. Genome-wide genetic diversity of Holstein Friesian cattle reveals new insights into Australian global population variability, including impact of selection. Animal Genetics 38, 7–14.

Zhao, H., Nuttleton, D., Soller, M., Dekkers, J.C.M., 2005. Evaluation of linkage disequilibrium measures between multi-allelic markers as predictors of linkage disequilibrium between markers and QTL. Genetical Research 80, 77–97.

Chapter 5

Genome-Wide Selection (GWS)

Marcos Deon Vilela de Resende,[a,b] **Fabyano Fonseca e Silva,**[b]
Márcio Fernando R. Resende Júnior,[c] **and Camila Ferreira Azevedo**[b]
[a]*Embrapa Florestas, Colombo, Brazil,* [b]*Federal University of Viçosa, Viçosa, Brazil,* [c]*University of Florida, Gainsville, FL, USA*

INTRODUCTION

There are four types of selection that use molecular markers: (1) gene-assisted selection (GAS), which is based on functional mutations and genes with known effects, or, in other words, the markers are the genes themselves; (2) marker-assisted selection using markers in linkage equilibrium with quantitative trait loci (QTLs) in the population (LE-MAS) but in linkage disequilibrium within families and crossings; (3) marker-assisted selection using markers in linkage disequilibrium at the population level (LD-MAS); and (4) genomic selection (GS) or genome-wide selection (GWS), which is based on thousands of markers in population-level linkage disequilibrium with all the QTLs for a polygenic trait. In GWS, there is no need to use phenotypic information in the selection population or to know and detect individual significant QTLs at arbitrary levels of significance. Phenotypes are only used for the discovery population to estimate the effects of various loci through markers. Of the four selection types, GWS tends to be the most efficient. Therefore, the present chapter will only discuss this method.

GENOME-WIDE SELECTION (GWS)

Genetic selection has been performed using the BLUP (best linear unbiased prediction) procedure (in its frequentist and Bayesian versions) with phenotypic data assessed in the field. The first attempt to increase the efficiency of BLUP based on phenotypic data was described by Lande and Thompson (1990), using molecular marker-assisted selection (MAS). MAS simultaneously uses phenotypic data and data from molecular markers in close linkage with some QTL. Generally, the marker data are used as covariables (fixed effects) when explaining phenotypic values of individuals under assessment or as random effects included in the phenotype model (Fernando and Grossman, 1989).

Biotechnology Applied to Plant Breeding. http://dx.doi.org/10.1016/B978-0-12-418672-9.00005-2

Such markers are chosen as determinants of the effects of QTLs after statistical modeling subject to type II errors (probability of accepting a false hypothesis or, in other words, to take as correct a false hypothesis of the absence of effects).

Selection based on MAS has the following features: it requires establishing marker-QTL associations (linkage analysis) for each family being studied or, in other words, those associations that are useful for selection only within each mapped family; to be useful, it must explain a large portion of the genetic variation for a quantitative trait that is controlled by many loci with small effects. This finding is not observed in practice, precisely because of the polygenic nature of quantitative traits and the great influence of the environment on them. These issues lead to the detection of only a few QTLs with large effects, and these are not sufficient to fully explain the genetic variation; MAS is only considerably better than selection based on phenotypic data when the studied family has been genotyped for a very large number of individuals (on the order of 500 or more). Due to these features, the implementation of MAS has been limited, and gains in efficiency have been very small (Dekkers, 2004).

The attraction of molecular genetics for the benefit of applied genetic improvement is the direct use of DNA information for selection to allow high selection efficiency, highly rapid achievement of genetic gains from selection, and low cost compared with traditional selection based on phenotypic data. Thus, Meuwissen et al. (2001) proposed a novel selection method called genomic selection (GS) or genome-wide selection (GWS). GWS can be applied to all families being studied in genetic improvement programs, features high selection accuracy for selection based only on markers (after their genetic effects have been estimated from phenotypic data in a sample of the selection population), does not require prior knowledge of QTL positions (map), and is not subject to type II errors related to selection of markers in linkage with QTLs.

This method remained relatively unknown in the scientific community for nearly 6 years because the available molecular markers at the time were costly and restricted. The recent development of low-cost SNP (*single nucleotide polymorphism*) markers has made the method more attractive (Meuwissen, 2007; Goddard and Hayes, 2007; Resende, 2007; Bernardo and Yu, 2007). GWS allows the prediction of genetic values and is excellent for traits with low heritability.

QTL analysis is based on the detection, mapping, and use of QTLs that strongly influence the traits under selection (MAS). It emphasizes the determination of the number, position, and effects of marked QTLs. GWS is defined as the simultaneous selection of hundreds or thousands of markers, which densely cover the genome such that all genes for a quantitative trait are in linkage disequilibrium with at least one portion of the markers. These markers in linkage disequilibrium with the QTLs, with both small and large effects, will explain nearly all of the genetic variation for a quantitative trait. The number of SNPs is so great that the probability of finding a QTL in linkage disequilibrium with at least one marker is very high. This feature is important because only the markers in linkage disequilibrium with QTLs are useful for determining phenotypes and explaining genetic variation. The effects of markers are estimated from a sample of individuals from the various families. Thus, the impact of certain specific families (with particular patterns of linkage disequilibrium) on the estimated marker effects is minimized. It should be noted that the genetic effects of the markers will be estimated from a sample of at least 1000 genotyped and phenotyped individuals; thus, it is based on at least 1000 experimental replicates for each locus. Therefore, even if the heritability of each effective marker (those that accurately identify one of the polygenes) is very low, the heritability will become high over 1000 replicates. Thus, the effect of the environment will be minimized by using a very high number of replicates. The same philosophy is applied for assessing and selecting quantitative traits based on phenotypes in field experiments.

GWS is wide because it acts on the entire genome to capture all of the genes that affect a quantitative trait. This feature is accomplished without the need for prior identification of markers with significant effects or mapping QTLs, as in the case of MAS. Genomic values associated with each marker or allele are used to provide the total genetic value for each individual. There is a fundamental difference in predicting traditional genetic values and predicting genomic values. For the former, phenotypic information is used to infer the effects of individual genotypes. For the latter, genotypic information (genotypes for the marker alleles) is used to infer future phenotypic values (or predicted genomic values) for individuals. In other words, the traditional methods use the phenotype to infer the effect of genotype, and GWS uses the genotype with the genetic effect previously estimated from a population sample to infer the phenotype to be expressed in the candidates for selection.

Marker effects are not the same in different studies and environments. In GWS, the genetic effects of markers are estimated for each trait of interest and used in the selection in each population for improvement in a particular environment. Estimation models that include the genotypes × environments interaction can also be used to test the possibility of obtaining valid estimates for a set of environments. However, the result will depend on the size of the interaction involving all environments.

GWS can be based on the use of the following: (1) only markers; (2) haplotypes or ranges defined by two markers; or (3) haplotypes defined by more than two markers, including covariance between haplotypes due to linkage. According to Callus et al. (2008), these three approaches do not differ for traits with low heritability (<10%). Solberg et al. (2006) demonstrated that GWS can be efficiently applied using directly the markers or, in other words, with direct prediction of marker effects without defining haplotypes. Those authors also report that this feature is advantageous because there is no need to estimate the linkage phase between markers, for which the estimates carry errors. Not only can SNPs be used in GWS, microsatellite markers can also be used. Solberg et al. (2006) claim that the use of SNPs requires four- to five-fold greater marker density than the use of microsatellites. This requirement is due to the bi-allelic (bi-nucleotidic) nature of SNPs and the multi-allelic nature of microsatellites. Such markers are efficient because they are codominant, multi-allelic, and abundant and are highly transferrable between individuals and species. DArT (*diversity array technology*) markers constitute another category that is well suited for GWS. These markers allow broad sampling of the genome without the need for prior knowledge of DNA sequences.

The conceptual development of GWS coincides with the technology associated with SNPs, which is accurate and relatively affordable. GWS uses associations between many SNP markers throughout the genome together with phenotypes and takes advantage of linkage disequilibrium between markers and QTLs in close linkage. The predictions derived from phenotypes and SNP genotypes with high density in a generation are thus used to obtain genomic values (GVs) for individuals in any subsequent generation based on their own genotypic markers.

When linkage disequilibrium between markers is incomplete, the joint allele frequencies for two loci can change markedly across generations, thereby leading to changes in haplotypes. In this case, the marker effects would need to be estimated again to maintain the accuracy of GWS for various generations (Dekkers, 2007). In the case of complete linkage disequilibrium, the estimated effects remain constant across different families and generations within the same environment.

ACCURACY OF GWS

The accuracy $(r_{q\hat{q}})$ of GWS depends on the proportion (r_{mq}^2) of genetic variation explained by the markers and the accuracy $(r_{m\hat{m}})$ of the prediction for effects of the markers or haplotypes that are in linkage disequilibrium with the QTLs, according to the formula $r_{q\hat{q}} = (r_{m\hat{m}}^2 r_{mq}^2)^{1/2}$. The parameter r_{mq}^2 depends on the marker density and the extent and pattern of linkage disequilibrium present in the population. The parameter $r_{m\hat{m}}$ depends on the amount and precision of data available to estimate the marker effects; it also depends on the efficiency of the strategy and statistical methods used for prediction.

Resende (2008) and Resende et al. (2008) offer an approach to calculate the expected accuracy of GWS, which was used by Grattapaglia and Resende (2010). The expected accuracy is given by $r_{q\hat{q}} = (r_{mq}^2 r_{m\hat{m}}^2)^{1/2} = \sqrt{r_{mq}^2 (Nh_m^2)/[1 + (N-1)h_m^2]}$. With the fitting of the residual polygenic effects in the model, $h_m^2 = (h^2 r_{mq}^2/n_Q)/(h^2 r_{mq}^2 + (1-h^2))$. Without such adjustment, $h_m^2 = (h^2 r_{mq}^2/n_Q)$

and $r_{q\hat{q}} = \sqrt{r_{mq}^2 (Nh^2 r_{mq}^2/n_Q)/[1 + (N-1)h^2 r_{mq}^2/n_Q]}$. The magnitude of linkage disequilibrium is calculated as follows (Sved, 1971):

$$r_{mq}^2 = E(r^2) = \frac{1}{4N_e L + 1}$$

The expected value for r^2, which measures the magnitude of linkage disequilibrium, depends on the effective population size (Ne) and the recombination frequency (a function of the size L of the chromosomal segment). The other quantities are the trait heritability (h^2), marker locus heritability (h_m^2), loci number (n_Q), and genotyped and phenotyped individuals (N).

For $Ne = 10$ and a distance of 1 cM between loci, the r^2 is expected to be 0.71. For the same distance between loci and Ne values of 20 and 30, the r^2 is expected to be 0.56 and 0.45, respectively. With twice as many markers and a distance of 0.5 cM between them, the expected r^2 are as follows: (1) $Ne = 10$; $r^2 = 0.83$; (2) $Ne = 20$; $r^2 = 0.71$; (3) $Ne = 30$; $r^2 = 0.63$. In eucalyptus, a species with a genome of 1.300 cM, at $Ne = 20$ with 1300 markers spaced at 1 cM, the r^2 is 0.56. Twice as many markers (2600) and a spacing of 0.5 cM between markers yield an expected r^2 of 0.71. However, 2600 would be the minimum number of markers for applying GWS in eucalyptus. In that case, with $N = 1000$ genotyped and phenotyped individuals, $n_Q = 100$ loci and h^2 at 30%, applying the accuracy formula would yield an accuracy of 70%, which is very interesting from a practical perspective.

Daetwyler et al. (2008) assumed residual variance $\sigma_e^2 = 1$ and $r_{mq}^2 = 1$,

and obtained $r_{m\hat{m}} = \sqrt{(Nh^2/n_Q)/[1 + (Nh^2/n_Q)]} = \sqrt{(\omega h^2)/[1 + (\omega h^2)]}$. An

important quantity is $\omega = N/n_Q$, which equals the number of individuals N used to estimate the effect of each locus in the estimation population. Resende (2008) obtained a more general expression, not assuming $\sigma_e^2 = 1$ and $r_{mq}^2 = 1$ and therefore keeping this last element in the formula.

In summary, the accuracy of GWS depends on five factors: (1) heritability of the trait; (2) number of loci regulating the trait and the distribution of their effects; (3) number of individuals in the discovery population; (4) effective population size; and (5) spacing between markers, which depends on their number and the genome size. The first two factors are beyond the breeder's control. The latter three factors can be modified by the breeder to increase the accuracy of GWS.

Table 5.1 shows the results for selection accuracy of GWS for a trait regulated by 100 loci with an individual narrow-sense heritability of 0.30. It is evident that for a eucalyptus population with an effective size of 20 ($r_{mq}^2 = 0.7$), the expected selection accuracy using GWS is 0.79 for a sample size of $N = 4000$ individuals. The expected accuracy is higher than the maximum accuracy (0.70) for individual selection by the traditional BLUP in mature trees. This finding confirms the great potential for GWS.

TABLE 5.1 Increase in the Accuracy* of GWS with the Increase of the Population Size. Trait Controlled by 100 Loci with Individual Narrow-Sense Heritability for Estimation of 0.30

Number of Individuals	$r^2_{mq} = 0.1$	$r^2_{mq} = 0.3$	$r^2_{mq} = 0.5$	$r^2_{mq} = 0.7$	$r^2_{mq} = 0.9$
100	0.06	0.18	0.27	0.36	0.44
200	0.09	0.24	0.36	0.47	0.57
500	0.13	0.33	0.48	0.61	0.72
1000	0.17	0.40	0.57	**0.70**	0.81
2000	0.21	0.46	0.62	0.76	0.87
4000	0.25	0.50	0.66	**0.79**	0.91
8000	0.28	0.52	0.68	0.81	0.93

*Maximum accuracy of individual selection for traditional BLUP in adults = 0.70.

TABLE 5.2 Efficiency of GWS per Unit Time

Phenotypic Accuracy (PA)	Genomic Accuracy (GA)	Phenotypic Time (PT)	Genomic Time (GT)	Efficiency (GA*PT)/(PA*GT)	Superiority %
0.70	0.79	4	4	1.13	13
0.70	0.79	4	3	1.50	50
0.70	0.79	4	2	2.26	126
0.70	0.79	4	1	4.51	351
0.70	0.79	4	0.5	9.03	803

It is evident that additional gains can be achieved by reducing the interval between selection cycles. For example, gains on the order of 126% can be obtained by reducing the interval from 4 to 2 years (Table 5.2).

ESTIMATION, VALIDATION, AND SELECTION POPULATIONS

In practice, three populations must be defined for GWS: estimation, validation, and selection populations (Goddard and Hayes, 2007; Meuwissen, 2007). These populations may be as follows: (1) physically distinct (three different

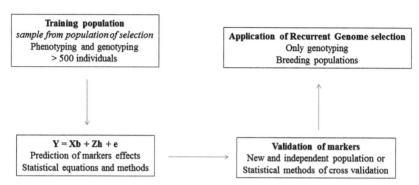

FIGURE 5.1　Schematic application of GWS in a genetic improvement program. (Resende et al., 2010)

populations); (2) with two functions simultaneously (only one population used for estimation and validation); or (3) with three functions simultaneously (only one population used for estimation, validation, and selection). Generally, strategies (1) and (2) are more commonly used. Figure 5.1 illustrates strategy (2).

Estimation population. The estimation population is also called the discovery, training, or reference population. This dataset includes a large number of markers assessed in a moderate number of individuals (1000 to 2000, depending on the desired accuracy), which should have their phenotypes assessed in terms of the various traits of interest. Equations for predicting genomic values (random multiple regression) are obtained for each trait of interest. These equations associate each marker or interval with its effect (predicted by RR-BLUP) on the trait of interest. The markers that explain the loci regulating the traits are identified in this population, and the marker effects are estimated.

Validation population. When physically separate from the estimation population, this dataset is smaller than the discovery population and includes individuals assessed for SNP markers and the various traits of interest. The equations for predicting genomic values are tested to verify their accuracy for this independent sample. To calculate the accuracy, genomic values are predicted (using the estimated effects from the estimation population) and subjected to correlation analysis with the observed phenotypic values. As the validation sample was not involved in predicting the marker effects, the errors from genomic values and phenotypic values are independent. Correlations between these values are predominantly genetic in nature and equivalent to the predictive ability ($r_{y\hat{y}}$) of GWS in estimating phenotypes, which is given by the accuracy of selection itself ($r_{q\hat{q}}$) multiplied by the square root of the heritability (h), or $r_{y\hat{y}} = r_{q\hat{q}}h$. Thus, to estimate the accuracy, one should obtain $r_{q\hat{q}} = r_{y\hat{y}}/h$. This method is valid when using raw phenotypic values to calculate correlation. When using genotypic values predicted based on phenotypes instead of raw phenotypic values, the heritability should be replaced by the reliability of the prediction. In general, strategy (2) is adopted according to a Jackknife scheme for cross validation, which will be described later. According to Meuwissen (2007), when dozens to hundreds of thousands of haplotypes are estimated, there is a risk of overparameterization; in other words, errors in the data being explained by the marker effects. Cross validation is therefore extremely important to address this problem.

Selection population. This dataset contains only the markers assessed in the candidates for selection. The phenotypes for this population do not need to be assessed. The prediction equations derived from the estimation population are therefore used to predict the GVs or future phenotypes of the candidates for selection. The associated selection accuracy is that calculated for the validation population.

The following strategy and analysis sequence involving the estimation and validation populations is recommended: in association with cross validation calculate the predicted GV using all markers, and calculate the correlation $r_{VGG,y}$ between GV and y; rank the markers by greatest modulus of the estimated marker effects; generate files with subsets of markers with the greatest moduli of the estimated marker effects (100, 250, 500, 1000, 1500, 2000, ...); analyze all of those files, calculate the correlations $r_{VGG,y}$, and then select the optimal file that maximizes $r_{VGG,y}$; perform a validation on that optimal file with $k = 2$ for the Jackknife process described below; validate the files smaller than the optimum and one larger to look for trends.

When estimating a parameter θ from a sample or dataset with n observations, the Jackknife procedure for estimating the variance of estimator $\hat{\theta}$ involves omitting each of the n observations, one per resampling. The generalized Jackknife method is based on dividing the set of N sample points into g groups of size k, such that $N = gk$. In general, k is set at 1 but can be as large as $N/2$. The estimator $\hat{\theta}_i$ is that which is based on samples of size $(g-1)k$, where the ith group of size k was removed. For $k = 1$, $N = g$ and $(g-1)k = g-1 = N-1$, such that $\hat{\theta}_i$ is the sample for which observation i was omitted (Resende, 2008). Validations with $k = 1$ and $k = 2$ tend to yield the same accuracies for the validation population. Thus, there is no need to use $k = 1$, and larger values are sufficient for cross validation.

Correlation and Regression among Predicted Genetic Values and Phenotypes in the Validation Population

The correlation and regression coefficients involving observed and predicted values are practical measures of the ability of the methods to make predictions that are accurate and unbiased, respectively. Correlation provides the predictive ability, which is equivalent to the product of accuracy and the square root of heritability. The regression coefficient is algebraically equal to 1. Regression coefficients less than 1 indicate that the genetic values are overestimated and exhibit greater variability than expected; coefficients greater than 1 indicate that the estimated genetic values exhibit less than expected variability. Lack of bias is important when selection involves individuals from many generations using estimated marker effects from a single generation. Regression coefficients near 1 indicate that the assessments are unbiased and effectively predict the actual magnitudes of differences between the individuals assessed.

The expected value of the regression coefficient is 1, and this case indicates an unbiased prediction. Thus, the regression coefficient can also be used to estimate the heritability captured by the markers. Various heritability values are assessed, and that which provides a regression equal to 1 should be selected as the best estimate. If the regression yields a result less than 1, the magnitude of the assessed heritability value was too high and should be reduced until converging the regression coefficient on 1. If the regression yields a result greater than 1, the magnitude of the assessed heritability value was too low and should be increased until converging on the regression 1.

Statistical Methods in Genome-Wide Selection

In the context of MAS and genomic prediction, the method of least squares (LS) has serious drawbacks. According to Gianola et al. (2003), the selection index (calculated as the regression involving molecular scores) presented by Lande and Thompson (1990) for MAS fails when formulated vectorially. This failure occurs because the covariance matrix for the molecular scores is singular, as the distribution of fitted regression values is defined only in p-dimensional space (number of covariables) and not in n-dimensional space (number of individuals with molecular scores). Therefore, the selection index leads to an infinite number of solutions.

Another difficulty arises when the number of markers is equal to or greater than the number of genotyped individuals. In this case, the collinearity of the predictor variables causes parametric identification problems, and some type of dimensional reduction, such as singular value decomposition, should be used. Another problem is the very inadmissibility (cannot provide the minimum mean square error) of LS estimators, a result that collapses estimates by LS and generalized least squares (GLS). Thus, the LS method is not recommended for MAS and GWS analyses. In summary, the LS method is inefficient because: it is impossible to simultaneously estimate all effects when the number of effects to be estimated is greater than the number of data points, estimating one effect at a time and testing its significance leads to overestimation of significant effects, and the accuracy of the method is low; only QTLs with large effects will be detected and used; and, consequently, not all of the genetic variation will be captured by the markers. The LS method assumes the *a priori* QTL distribution, with an infinitely large variance that disagrees with the known total genetic variance.

Because the number of markers in GWS is greater than the number of individuals, there is a lack of degrees of freedom to estimate the effects of all markers. A solution for this problem is to use the method of ridge regression (RR) (Whittaker et al., 2000) or to take the marker effects as random instead of fixed. Fitting random effects does not expend degrees of freedom; therefore, the effects of all markers can be simultaneously estimated. This method leads to RR-BLUP, which takes into account the effects of QTL with normal distributions and equal variance through the chromosomal segments. The distribution of QTL effects is known for few traits and species. In dairy cattle, for example, Goddard and Hayes (2007) reported the presence of 150 QTLs for the trait milk production and estimated the distribution of their effects as approximately exponential.

Generic Description of GWS Methods

The ideal method for GWS should exhibit three features: it should (1) accommodate the genetic architecture of the trait in terms of genes with small and large effects and their distributions; (2) regularize the estimation process for the presence of multicollinearity and a large number of markers by using *shrinkage* estimators; and (3) select the covariables (markers) that affect the analyzed trait.

The main problem for GWS is estimating the large number of effects from a limited number of observations, in addition to collinearities resulting from linkage disequilibrium between markers. The shrinkage estimators adequately address this issue by treating the marker effects as random variables and estimating them simultaneously (Resende et al., 2008).

The main GWS methods can be divided into three large classes: explicit regression, implicit regression, and regression with dimensional reduction (Resende et al., 2011). The first class features methods such as RR-BLUP, LASSO (*Least Absolute Shrinkage and Selection Operator*), elastic net (EN), BayesA and BayesB, among others. Implicit regression includes the methods of RKHS (*Reproducing Kernel Hilbert Spaces*, which is a semiparametric method, similar to the neural nets method) (Gianola and Campos, 2009) and nonparametric kernel regression via generalized additive models (Gianola et al., 2006). The methods for regression with dimensional reduction feature partial least squares, independent components (Azevedo et al., 2013), and principal components (Table 5.3).

Methods for explicit regression are divided into two groups: (1) methods for penalized estimation (RR-BLUP, LASSO, EN, RR-BLUP-Het) and (2) methods for Bayesian estimation (BayesA, BayesB, Fast BayesB, BayesCπ, BayesDπ, BayesR, BayesRS, BLASSO, IBLASSO, and others) (Table 5.3). BayesR (Erbe et al., 2012) and BayesRS (Brondum et al., 2012) are new Bayesian methods. The penalized estimators are obtained as a solution to an optimization problem, where the objective function (function for which the value is minimized or maximized, depending on the problem and objective) is defined by the balance between precision of fit (residual sum of squares) and model complexity (penalization component). The methods for penalized estimation differ in the

TABLE 5.3 Classification of GWS Methods

Class	Family	Method	Attributes
Explicit regression	Methods of penalized estimation (linear regression)	RR-BLUP/GWS	Regularization
			Homogeneous genetic architecture
			Indirect covariable selection
		LASSO	Regularization
			Homogeneous genetic architecture
			Direct covariable selection
		EN	Regularization
			Homogeneous genetic architecture
			Direct covariable selection
		RR-BLUP-Het/GWS	Regularization
			Flexible genetic architecture
			Indirect covariable selection
	Methods for Bayesian estimation (nonlinear regression)	BayesA	Regularization
			Flexible genetic architecture
			Indirect covariable selection
		BayesB	Regularization
			Flexible genetic architecture
			Guided covariable selection
		Fast BayesB	Regularization
			Flexible genetic architecture
			Guided covariable selection

(Continued)

TABLE 5.3 Continued

Class	Family	Method	Attributes
		BayesCπ	Regularization
			Homogeneous genetic architecture
			Direct covariable selection
		BayesDπ	Regularization
			Flexible genetic architecture
			Direct covariable selection
		BLASSO	Regularization
			Flexible genetic architecture
			Direct covariable selection
		IBLASSO	Regularization
			Flexible genetic architecture
			Direct covariable selection
Implicit regression		Kernel Regression	
		RKHS	
		Neural networks	
Regression with dimensional reduction		Partial least squares	
		Principal components	
		Independent components	

penalization functions used, which produce different degrees of shrinkage. This shrinkage prevents overparameterization and can lead to a reduction in the mean squared error of the estimation.

Bayesian methods are associated with systems of nonlinear equations, and nonlinear predictions can be better when the QTL effects are not normally distributed due to the presence of genes with major effects. The linear predictions associated with RR-BLUP assume that all markers with the same allele frequency contribute equally to the genetic variation (lack of genes with major effects). In Bayesian estimation, the shrinkage of effect estimates for the model is controlled by the *a priori* distribution assumed for those effects. Different distributions produce different shrinkages. Methods for penalized and Bayesian estimation may have (BayesB, Fast BayesB, BayesCπ, BayesDπ, LASSO, BLASSO, IBLASSO) or lack (RR-BLUP, EN, RR-BLUP-Het, BayesA) direct covariable selection. Bayesian methods are better when the distribution of QTL effects is leptokurtic (positive kurtosis) due to the presence of genes with large effects. When the QTL effects are normally distributed, the RR-BLUP method is equally efficient.

Comparisons among methods for predicting genomic values have been made. Meuwissen et al. (2001) concluded that the BayesB method is theoretically best; this method proved slightly better than RR-BLUP. However, the author simulated genotypic data with the same *a priori* distribution used for estimation. This approach yielded greater accuracy for this method, although such accuracy is unattainable in practice if the actual distribution associated with genetic effects differs from the *a priori* distribution assumed for analysis. In general, there is no method that is best under all circumstances because each method may yield significantly different results depending on the population structure and nature of the trait. However, the results obtained by Guo et al. (2012) indicate that the RR-BLUP method is equal to or better than the others for most applications in plants and is also easier to apply. Thus, this chapter will deal exclusively with this procedure.

RR-BLUP Method

The RR-BLUP method uses BLUP predictors, and the marker effects are fit as covariables of random effects. In other words, the phenotypes are regressed onto the covariables. Because they are covariables (quantitative scale) and not classificatory variables (qualitative scale of 0 and 1), different incidence matrices and therefore different computational algorithms are produced from traditional BLUP. The most appropriate name is random regression of the BLUP type (RR-BLUP) applied to genome-wide selection (RR-BLUP/GWS). The technique of random regression is a generalization of the ridge regression (RR).

The estimators associated with random regression and RR promote shrinkage dictated by a function of the quantity λ. When λ is unknown, its arbitrary selection should lead to the RR method. If the parameter for regression is associated with $\lambda = \sigma_e^2/\sigma_{ai}^2 = \sigma_e^2/(\sigma_a^2/n_Q)$, RR-BLUP is used for the effect of chromosomal segment i, where σ_{ai}^2 is the additive genetic variance associated with the locus or segment i and σ_a^2 and σ_e^2 are the additive genetic variance of the trait and the residual variance, respectively. The quantity n_Q is unknown *a priori* but can be inferred as described below. The penalization parameter λ can also be determined iteratively or by fine tuning, selecting the parameter that maximizes correlation between the phenotypic value and the predicted genetic value in the cross validation. Whittaker et al. (2000) and Meuwissen et al. (2001) were pioneers for proposing simultaneous prediction of marker effects without the use of significance testing for individual markers. This approach contrasts with the MAS method proposed by Lande and Thompson (1990) (Table 5.4).

It is evident that the innovation of Meuwissen et al. (2001) of emphasizing the use of linkage disequilibrium in the population and not just within the family was based on conceptual and statistical terms. Additionally, this approach did not use significance tests for markers. In any case, the authors' greatest achievement was to demonstrate, by simulation, that GWS can be efficient in practice.

That GWS does not emphasize significance testing for marker selection is an important feature for distinguishing this method from GWAS (*genome-wide association studies*) (Chapter 4). The objective of the latter is to identify loci with significant effects on the trait (hypothesis testing where H_0 is a null effect of the locus). GWAS suffer high false-negative rates due to the use of stringent significance levels that are accepted to avoid false-positives. GWS is equivalent to GWAS being applied to all loci simultaneously based on estimation and prediction rather than hypothesis testing. Thus, GWS can explain much more of the genetic variability and avoid the so-called missing or lost heritability that is typical of linkage and association studies.

The distinction between fixed, ridge, and random regressions in a model using only phenotypes is the penalization parameter λ^*, which is given by $\lambda^* = (1 - h^2)/h^2$. Small values for λ^* are sufficient to reduce the effect of multicollinearity among the columns in matrix $X'X$, which is approximately singular. A λ^* of 0 ($h^2 = 1$) is characteristic of fixed regression. Small values for λ^* (0.01 to 1) are characteristic of ridge regression, whereas large values for λ^* (greater than 0.1) are characteristic of random regression.

Prediction by RR-BLUP is described below based on Resende (2007, 2008). The following linear mixed model is fit to estimate the marker effects: $y = Wb + Xm + e$, where y is the vector of phenotypic observations, b is the vector of fixed effects, m is the vector of random marker effects, and e is the vector of random residuals. W and X are the incidence matrices for b and m. The incidence matrix X contains functions of the values 0, 1, and 2 for the number of alleles for the marker (or the supposed QTL) in a diploid individual. A similar coding method is to use the values -1, 0, and 1. Genomic mixed-model equations for predicting m by the RR-BLUP method are equivalent to:

$$
\begin{bmatrix} W'W & W'X \\ X'W & X'X + I\dfrac{\sigma_e^2}{(\sigma_a^2/n_Q)} \end{bmatrix} \begin{bmatrix} \hat{b} \\ \hat{m} \end{bmatrix} = \begin{bmatrix} W'y \\ X'y \end{bmatrix}
$$

The total additive genomic value for an individual j is given by $VGG_j = \hat{y}_j = \sum_i X_i \hat{m}_i$, where X_i is equal to 0, 1, or 2 for genotypes aa, Aa, and AA, respectively, for bi-allelic and codominant markers, such as SNPs.

The prediction equations presented above assume *a priori* that all loci explain equal amounts of the genetic variation. Thus, the genetic variation

TABLE 5.4 Comparison of the Three Proposals for Marker-Assisted Selection

Authors	Method	Population	Number of Markers (m)	Significance Test	Extension of the Bayesian Approach
Lande and Thompson (1990)	MAS – Index of Multiple Regression Selection	Within the family or crossing	Much smaller than the size of the crossing (N): $m \ll N$	Yes	No
Whittaker et al. (2000)	MAS – Ridge Regression	Within the family or crossing	Greater or equal to the size of the crossing (N): $m \geq N$	Yes	Yes
Meuwissen et al. (2001)	GWS – RR-BLUP	Entire population	Much larger than the size of the crossing (N): $m \gg N$	No	Yes

explained by each locus is given by σ_a^2/n_Q, where σ_a^2 is the total genetic variation and n_Q is the number of loci (when each locus is perfectly marked by a single marker), which can be given by $n_Q = 2\sum_i^n p_i(1-p_i)$, where p_i is the frequency of the allele of the kind A in the locus i. The genetic variation σ_a^2 can be estimated by REML on the phenotypic data in the traditional manner or by the very variation among markers or QTL chromosomal segments, as described below.

There is no need to use the kinship matrix with the RR-BLUP method. The kinship matrix based on pedigree used for traditional BLUP is replaced by a kinship matrix estimated by the markers. This kinship matrix is a function of the $X'X$ present in the equations of the mixed model. This procedure is better than using the pedigree because it effectively captures the kinship matrix produced for each trait and not an average kinship matrix associated with the pedigree. For example, the additive genetic correlation between two full sibs, based on pedigree, is 0.5. However, the markers can indicate that the actual value is a fraction between 0 and 1. The value of 0.5 is expected on average. However, the correlation can be 0, 0.5, or 1 at each locus depending on the number of identical alleles shared by both sibs.

GWS improves the accuracy of the estimate $\hat{a}_d$, which refers to the effects of Mendelian segregation within families and is the method that appropriately exploits the segregation of Mendelian sampling that occurs during gamete formation. As GWS directly assesses the DNA associated (via markers) with each locus for the entire polygenic trait, it directly assesses each segregation at the individual level and not at the average level. By directly assessing the genotype of offspring it is possible to recognize each segregation. According to Goddard and Hayes (2007), under the infinitesimal model with a great number of loci of small effects, GWS predicts the genetic values more accurately than traditional BLUP based on pedigree and phenotypic data. GWS places more emphasis on the term $\hat{a}_d$, which refers to Mendelian segregation, and places greater weight on this component than does traditional BLUP. This difference leads to the selection of less related individuals than in BLUP, which reduces the increase of inbreeding in the population.

The kinship matrix can also be computed separately and incorporated into the equations of the mixed model for traditional BLUP, according to model (3) described below, producing the genomic BLUP (*G-BLUP*) method. In this case, the matrix is given by $G = (X^*X^{*\prime})/[2\sum_i^n p_i(1 - p_i)]$, where p_i is the frequency of one of the alleles of locus i and X^* is the matrix X corrected for the mean at each locus ($2p_i$). To guarantee that A is a defined positive matrix, one can obtain $G^p = G + 10^{-6} I$, where I is an identity matrix. The genomic inbreeding coefficient for individual i is given by $G_{ii} - 1$. Another way to obtain G is by $G = X^*DX^{*\prime}$, where D is diagonal with D_{ii} given by $D_{ii} = 1/\{n[2p_i(1 - p_i)]\}$, where n is the number of markers.

The diagonal of matrix XX' describes the kinship of an individual with himself, and the elements off the diagonal show the number of alleles shared by kin. The Wright correlation between kin can be obtained by dividing those off-diagonal elements by the product of the square roots of the respective elements from the diagonal. The diagonal of matrix $X'X$ shows how many individuals inherited each allele, and elements off the diagonal indicate how many times two different alleles were inherited by the same individual (Van Raden, 2008). Using genomic methods, the concept of inbreeding at a neutral locus is not more valid because measures of kinship are considered at the loci of the very trait under selection. The traditional measures of inbreeding based on pedigree result in much more variable losses of diversity.

The prediction of genomic values by BLUP can be calculated using three equivalent methods (Van Raden, 2008):

1. By RR-BLUP for markers, as specified, where: $\hat{g} = X\hat{m} = X(X'R^{-1}X + I\lambda)^{-1}X'R^{-1}(y - W\hat{b})$, given that $\hat{m} = (X'R^{-1}X + I\lambda)^{-1}X'R^{-1}(y - W\hat{b})$. R is a diagonal matrix of weightings that take into account different reliabilities among individuals' evaluations. When the reliabilities are high and homogeneous (greater than 0.85), then $R = I$ and the system reduces to $\hat{m} = (X'X + I\lambda)^{-1}X'(y - W\hat{b})$.

2. By BLUP or the selection index (where G is genomic and b is estimated by generalized least squares, which is guaranteed when y contains deregressed genetic values),

where: $\hat{g} = G[G + R(\sigma_e^2/\sigma_g^2)]^{-1}(y - W\hat{b})$. If necessary, marker effects can be obtained from $\hat{m} = \{X'/[2\sum_i^n p_i(1 - p_i)]\}[G + R(\sigma_e^2/\sigma_g^2)]^{-1}(y - W\hat{b})$.

3. By the BLUP Equivalent Model (G-BLUP), where: $\hat{g} = [R^{-1} + G^{-1}(\sigma_e^2/\sigma_g^2)]^{-1}R^{-1}(y - W\hat{b})$.

The following elements are required to apply the RR-BLUP/GWS procedure: W, X, y, and $\lambda = \sigma_e^2/\sigma_{ai}^2 = \sigma_e^2/(\sigma_a^2/n_Q)$. Vector y corresponds to the corrected phenotypes, matrix X corresponds to the counts of allele numbers of the molecular markers, W is a known vector consisting of values of 1, and λ depends on the variance components (heritability or selection reliability) and the effective number of chromosomal segments, associated to n_Q. Each of these elements is described below, according to Resende et al. (2010).

Methods of Parameterizing the Genotypic Incidence Matrix

Parameterization 1

The incidence matrix X contains the values 0, 1, and 2 for the number of alleles for a marker (or supposed QTL) in a diploid individual and $2p$ for individuals with lost marker data. These values should be centered around 0 such that the effects of codominant markers are effects of allelic substitution with a mean of 0 in the population. In this case, assuming Hardy-Weinberg equilibrium, the additive genetic variation of the trait in the population is equal to $\sigma_a^2 = 2\sum_i^n p_i(1 - p_i)\sigma_m^2$. Thus, the values of X_i should be replaced by $0 - 2p$, $1 - 2p$, and $2 - 2p$, respectively, to obtain a variable with a mean of 0. Thus, with centralization, $n_Q = 2\sum_i^n p_i(1 - p_i)$ should be used for the RR-BLUP method, and the additive genetic effects of individuals are given by $\hat{a} = X\hat{m}$.

Additionally, the data for markers in matrix X can be standardized as follows for each matrix element X_i, corresponding to locus i:

$X_i = (0 - 2p_i)/(\text{Var}(X_i))^{1/2}$ if the individual is homozygous for the first allele (aa);

$X_i = (1 - 2p_i)/(\text{Var}(X_i))^{1/2}$ if the individual is heterozygous (Aa);

$X_i = (2 - 2p_i)/2(\text{Var}(X_i))^{1/2}$ if the individual is homozygous for the second allele at the marker locus (AA);

$X_i = 0$ if the individual has missing marker data.

The quantity p_i is the frequency of the second marker allele. Thus, the variance of X with X_i adjusted is equal to 1, obtaining a variable with a mean of 0 and unitary variance. Taking m as the marker effect in the population, the variance attributed to the marker is given by $\text{Var}(X_i m) = \text{Var}(X_i)\,\text{Var}(m)$. With the above transformation, $\text{Var}(X_i) = 1$; therefore, $\text{Var}(X_i m) = \text{Var}(m)$. In other words, the modeling of the variance of the marker effect is being done independently of its frequency. Thus, after centralization and normalization, $\sigma_a^2 = n\sigma_m^2$. As such, $n_Q = n$ should be used for the RR-BLUP method, and the additive genetic effects of individuals are given by $\hat{a} = X_i\hat{m}$.

Parameterization 2

In another parameterization, the incidence matrix X contains the values -1, 0, and 1 for the number of alleles of the marker (or the supposed QTL) in a diploid individual, or in other words, for the genotypes aa, Aa, and AA, respectively. For this parameterization, which is slightly inferior to the prior one (Legarra et al., 2011), the following equation should be used for the RR-BLUP method: $n_Q = 2\sum_i^n p_i(1 - p_i)$, and the additive genetic effect of individual j is given by

$$\hat{a}_j = \sum_i^n [I(x_{ij} = 1)(2p_i\hat{m}_i) + I(x_{ij} = 0)$$
$$(p_i\hat{m}_i - q_i\hat{m}_i) + I(x_{ij} = -1)(-2q_i\hat{m}_i)$$

Phenotype Correction

The phenotypes should be corrected for the effects of the environment and the progenitors (population structure). Thus, the genetic values should be predicted and subsequently deregressed and corrected for the effect of the progenitors. Values should be deregressed for three reasons: there cannot be two regressions, one based on pedigree and another based on markers; the matrix A based on pedigree is less precise than matrix $G = (XX')/n$ based on markers; and genes with a large effect present in one of the progenitors will have an influence. Additionally, the phenotypes should be corrected for the genetic effects of the progenitors, basically working with the effect of *deregressed Mendelian segregation* because the ideal datum for the training population should be the *actual generic merit of unrelated individuals* (Garrick et al., 2009). The effect of Mendelian segregation provides the following: analysis of the association of alleles for markers and QTLs, or, in other words, the linkage disequilibrium (LD) independent of genealogy.

Another explicit manner of partially accomplishing this is to consider the pedigree by fitting a^*, the vector of polygenic effects, using the model $y = Wbm + Ta^* + e$, where T is the incidence matrix for a^*. Without the correction mentioned above or fitting a^*, the markers may only capture the kinship (population structure) among individuals and not necessarily the linkage disequilibrium with the genes themselves. In that case, the accuracy of validation may be low in an independent sample (individuals from other families) of the population and individuals from other generations, in contrast to what would have been expected from GWS.

The procedure for obtaining the deregressed and corrected phenotypic values for genetic effects of the progenitors was described by Garrick et al. (2009) and Resende et al. (2010). For short-term GWS there is no need for such correction. Another method of performing that fit is by fitting the effects of progenitors as fixed effects (Vazquez et al., 2010). Fitting in this way captures the effects of progenitors from the individual genetic values and leaves only the effects of Mendelian segregation, which should be deregressed.

RELATIONSHIP BETWEEN GENETIC VARIANCE AND MARKER VARIANCE

The relationship between additive genetic variance and the variance of marker effects is essential for genomic prediction. It is assumed that $\mathrm{Var}(a_i) = \mathrm{Var}(X_i m_i) = \mathrm{Var}(X_i)\mathrm{Var}(m_i) = 2p_i(1 - p_i)\mathrm{Var}(m_i) = 2p_i(1 - p_i)m_i^2$ equals the genetic variance attributed to locus i. For various loci, the total additive genetic variance is given by $\sigma_a^2 = \sum_i^n 2p_i(1 - p_i)m_i^2$, which can also be expressed as $\sigma_a^2 = \sum_i^n U_i V_i$, where $U_i = 2p_i(1 - p_i)$ and $V_i = m_i^2$. The covariance between U and V, called C_{UV}, is given by $C_{UV} = (\sum_i^n U_i V_i)/n - (\sum_i^n U_i/n)(\sum_i^n V_i/n)$. Rearranging this expression gives $\sum_i^n U_i V_i = nC_{UV} + (\sum_i^n U_i)(\sum_i^n V_i/n)$, such that $\sigma_a^2 = \sum_i^n U_i V_i = nC_{UV} + [\sum_i^n 2p_i(1 - p_i)](\sum_i^n m_i^2)/n$. Given that $(\sum_i^n m_i^2)/n = \sigma_m^2$, then $\sigma_a^2 = [2\sum_i^n p_i(1 - p_i)\sigma_m^2] + nC_{UV}$.

Thus the variance among markers (σ_m^2) obtained by REML, the allele frequencies, and the marker effects predicted by BLUP can be used to obtain the total additive genetic variance. In some cases, C_{UV} tends toward 0. In other cases, the quantity m_i^2 is replaced by σ_m^2, because the expectation for m_i^2 is the variance of the marker effect; in other words, $E(m_i^2) = \sigma_m^2$. Therefore, many applications use $\sigma_a^2 = [2\sum_i^n p_i(1 - p_i)\sigma_m^2]$, and the variance among markers given by $\sigma_m^2 = (\sigma_a^2 - nC_{UV})/[2\sum_i^n p_i(1 - p_i)]$ is reduced to $\sigma_m^2 = \sigma_{ai}^2 = \sigma_a^2/[2\sum_i^n p_i(1 - p_i)]$.

The quantity $\lambda = \sigma_e^2/\sigma_{ai}^2 = \sigma_e^2/(\sigma_a^2/n_Q)$ is required for RR-BLUP/GWS prediction, where n_Q is the number of loci regulating the trait (assuming that each locus is perfectly marked), which is not known *a priori*. For $\sigma_{ai}^2 = \sigma_a^2/[2\sum_i^n p_i(1 - p_i)]$, n_Q can be taken as $[2\sum_i^n p_i(1 - p_i)]$. Alternatively, λ can be expressed as $\lambda = n_Q(1 - h^2)/h^2 = [2\sum_i^n p_i(1 - p_i)](1 - h^2)/h^2$. Thus, having h^2 and the allele frequencies at the marker loci, λ is obtained for use in the mixed-model equations.

The genetic variance and heritability (h^2) can be calculated using phenotypic data or using marker and phenotypic data, as described for the calculation of σ_a^2. When h^2 is used for RR-BLUP, it should be the adjusted heritability or from corrected data ($h_{aj}^2 = \sigma_a^2/\sigma_{yaj}^2$), where σ_{yaj} is the adjusted phenotypic variance. If y is corrected for the mean of the progenitors, the numerator of h_{aj}^2 should only contain the genetic variance attributed to Mendelian segregation. In other words, $h_{aj}^{2*} = (1/2)\sigma_a^2/\sigma_{yaj}^2$ or $h_{aj}^{2*} = (3/4)\sigma_a^2/\sigma_{yaj}^2$ when both progenitors (full sib families) or only one progenitor (half-sib families) are known, respectively. These heritabilities can also be expressed as a function of individual heritability h^2, using the expression $h_{aj}^{2*} = (1/2h^2)/(1/2h^2 + (1 - h^2))$ for full sib progenies and $h_{aj}^{2*} = (3/4h^2)/(3/4h^2 + (1 - h^2))$ for half-sib progenies. These formulas show that the denominator of h_{aj}^{2*} also contains only the genetic variance attributed to Mendelian segregation and not the total genetic variation. Another way of expressing h_{aj}^{2*} is to directly use the reliability or squared accuracy (r_i^{2*})

of individuals' evaluation. For calculating RR-BLUP and the accuracy of GWS, h_{aj}^{2*} can be taken as the mean of r_i^{2*} for the individuals being analyzed.

Initially analyzing the entire set of codominant markers in all phenotyped individuals (total estimation population) is recommended. This procedure seeks to identify the markers with the greatest effects in the modulus, with the aim of analyzing smaller subgroups of markers and determining how many and which markers maximize the accuracy of selection. The optimal number of markers is a tradeoff between informativity (greater accuracy because of the better capture of genes) and less precision (less accuracy because of the smaller sample size per estimated effect) with an increase in the number of markers. Subsequently, validation should be performed using only the subset of markers that maximize accuracy, taking n as the sum $[2\sum_i^n p_i(1-p_i)]$ in that subset of markers. Additionally, σ_m^2 and h^2 should be recalculated because h^2 can be smaller than previously calculated. However, the h^2 used to calculate the accuracy from the predictive ability, using $r_{q\hat{q}} = r_{y\hat{y}}/h$, should be the total h^2, estimated from the phenotypic data. This h^2 tends to be similar to the h^2 estimated using markers when the total number of markers used is large. This procedure (called RR-BLUP_B by Resende et al., 2010, 2012) is recommended, as it tends to produce greater accuracy, similar to that obtained by Bayesian methods, which perform the covariables selection automatically. Therefore, both approaches assume that many markers have no effect.

The increase or decrease in the accuracy of GWS via RR-BLUP is a compromise or tradeoff between increasing the amount of useful information by using more marker loci and reducing the effective sample size to estimate the effect of each locus; thus, fewer individuals per locus to be estimated (smaller N/n).

The lower number of markers explaining a large part of the genetic variation or the maximum possible accuracy is very interesting from a practical perspective. In this case, DNA arrangements with a low density of previously selected markers can be used in selection populations. In Australia, genomic values for dairy cattle can be accurately predicted using chips containing 1000 (offering 85% accuracy obtained with 42,500 SNPs) to 5000 (offering 95% accuracy obtained with 42,500 SNPs) evenly spaced SNPs (Moser et al., 2010). An alternative to using previously selected markers is using evenly spaced markers in greater number than those selected. This approach allows dealing with various traits and can lead to the generalized use of GWS in various species in different countries.

Example Using RR-BLUP/GWS

Consider the small example shown in Table 5.5, which involves the evaluation of five individuals for the diameter trait and genotyping for seven markers. The number of one of the alleles is given for each marker locus.

The genetic effects of the markers are obtained by solving

$$
\begin{bmatrix}
W'W & W'X \\
X'W & X'X + I\dfrac{\sigma_e^2}{(\sigma_a^2/n_Q)}
\end{bmatrix}
\begin{bmatrix}
\hat{b} \\
\hat{m}
\end{bmatrix}
=
\begin{bmatrix}
W'y \\
X'y
\end{bmatrix}
$$

TABLE 5.5 Example of Evaluation of Five Individuals for the Diameter Trait and Genotyping for Seven Markers

Individual	Diameter	Markers						
		1	2	3	4	5	6	7
1	9.87	2	0	0	0	2	0	0
2	14.48	1	1	0	0	1	1	0
3	8.91	0	2	0	0	0	0	2
4	14.64	1	0	1	0	1	0	0
5	9.55	1	0	0	1	1	1	0

The following matrices are obtained:

$$X = \begin{bmatrix} 2 & 0 & 0 & 0 & 2 & 0 & 0 \\ 1 & 1 & 0 & 0 & 1 & 1 & 0 \\ 0 & 2 & 0 & 0 & 0 & 0 & 2 \\ 1 & 0 & 1 & 0 & 1 & 0 & 0 \\ 1 & 0 & 0 & 1 & 1 & 1 & 0 \end{bmatrix} \quad y = \begin{bmatrix} 9.87 \\ 14.48 \\ 8.91 \\ 14.64 \\ 9.55 \end{bmatrix} ; \quad W = \begin{bmatrix} 1 \\ 1 \\ 1 \\ 1 \\ 1 \end{bmatrix}$$

Performing the multiplications and assuming that

$$\frac{\sigma_e^2}{\left(\sigma_a^2/n_Q\right)} = 1$$

gives

$$W'W = [5]; \quad W'X = \begin{bmatrix} 5 & 3 & 1 & 1 & 5 & 2 & 2 \end{bmatrix};$$

$$X'W = (W'X)' = \begin{bmatrix} 5 & 3 & 1 & 1 & 5 & 2 & 2 \end{bmatrix}'$$

$$X'X + I = \begin{bmatrix} 8 & 1 & 1 & 1 & 7 & 2 & 0 \\ 1 & 6 & 0 & 0 & 1 & 1 & 4 \\ 1 & 0 & 2 & 0 & 1 & 0 & 0 \\ 1 & 0 & 0 & 2 & 1 & 1 & 0 \\ 7 & 1 & 1 & 1 & 8 & 2 & 0 \\ 2 & 1 & 0 & 1 & 2 & 3 & 0 \\ 0 & 4 & 0 & 0 & 0 & 0 & 5 \end{bmatrix} \quad W'y = [57.45] \quad X'y = \begin{bmatrix} 58.4100 \\ 32.3000 \\ 14.6400 \\ 9.5500 \\ 58.4100 \\ 24.0300 \\ 17.8200 \end{bmatrix} .$$

Therefore:

$$
\begin{bmatrix} \hat{b} \\ \hat{m} \end{bmatrix} = \begin{bmatrix} 5 & 5 & 3 & 1 & 1 & 5 & 2 & 2 \\ 5 & 8 & 1 & 1 & 1 & 7 & 2 & 0 \\ 3 & 1 & 6 & 0 & 0 & 1 & 1 & 4 \\ 1 & 1 & 0 & 2 & 0 & 1 & 0 & 0 \\ 1 & 1 & 0 & 0 & 2 & 1 & 1 & 0 \\ 5 & 7 & 1 & 1 & 1 & 8 & 2 & 0 \\ 2 & 2 & 1 & 0 & 1 & 2 & 3 & 0 \\ 2 & 0 & 4 & 0 & 0 & 0 & 0 & 5 \end{bmatrix}^{-1} \begin{bmatrix} 57.4500 \\ 58.4100 \\ 32.3000 \\ 14.6400 \\ 9.5500 \\ 58.4100 \\ 24.0300 \\ 17.8200 \end{bmatrix}
$$

The results are

$$
\begin{bmatrix} \hat{b} \\ \hat{m} \end{bmatrix} = \begin{bmatrix} 12.4519 \\ -0.3526 \\ 0.2761 \\ 1.4467 \\ -1.3701 \\ -0.3526 \\ 0.5436 \\ -1.63765 \end{bmatrix},
$$

where 12.4519 is the general mean and the remaining values are the estimates of the marker effects.

The genomic value of individuals from a selection population can be obtained by $VGG_j = \hat{y}_j = \sum_i X_i \hat{m}_i$. In this case, the predictions are

$$
VGG = \begin{bmatrix} -1.4104 \\ 0.1145 \\ -2.7230 \\ 0.7415 \\ -1.5317 \end{bmatrix}
$$

Simultaneous Analysis of Genotyped and Nongenotyped Individuals via GBLUP

For the group of genotyped and phenotyped individuals, the following linear mixed model is fitted to estimate the additive genetic effects using phenotypic information (Resende, 2008; Resende et al., 2010): $y = Wb + Za + e$, where y is the vector of phenotypic observations, b is the vector of fixed effects, a is the vector of additive individual genetic effects (random), and e is the vector of random residuals. W and Z are the incidence matrices for b and a. Using phenotypic and marker information gives the equivalent model: $y = Wb + ZXm + e$, where m is the vector of random marker effects, X is the incidence matrix for m and $a = Xm$. The incidence matrix X contains the values 0, 1, and 2 for the number of alleles for the marker (or supposed QTL) in a diploid individual. Another equivalent way of coding is to use the values $-1, 0$, and 1.

The mixed-model equations for predicting a using the G-BLUP method are equivalent to:

$$\begin{bmatrix} W'W & W'Z \\ Z'W & Z'Z + G^{-1}\frac{\sigma_e^2}{\sigma_a^2} \end{bmatrix} \begin{bmatrix} \hat{b} \\ \hat{a} \end{bmatrix} = \begin{bmatrix} W'y \\ Z'y \end{bmatrix},$$

where $G = (XX')/k = (XX')/[2\sum_i^n p_i(1-p_i)]$ and $k = 2\sum_i^n p_i(1-p_i)$. Prior normalization of the elements in X (dividing them by $[2\sum_i^n p_i(1-p_i)]^{1/2}$) and centering the mean on 0 gives $G = XX'/n$. The scaling parameter $k = 2\sum_i^n p_i(1-p_i)$ assumes independence of the SNP effects. To bypass that assumption, Gianola et al. (2009) proposed the following scaling parameter:

$$k = \left((p_0 - q_0)^2 + 2\left(\left[\sum_i^n p_i(1-p_i) \right]/n \right) ((\alpha + \beta + 2)/(\alpha + \beta)) \right) n,$$

where $p_0 = \alpha/(\alpha + \beta)$, the expected allele frequency $q_0 = (1 - p_0)$, α and β are beta distribution parameters fitting the basic allele frequency, and n is the number of SNPs. The estimator for a can be summarized as:

$$[\hat{a}] = \left[Z'Z + G^{-1}\frac{\sigma_e^2}{\sigma_a^2} \right]^{-1} \left[Z'y \right]$$

For a global evaluation of the three classes of individuals in a single step, the same model $y = Wb + Za + e$ can be used with one alteration (replacing matrix G with matrix H) to the mixed-model equations, according to Misztal et al. (2009):

$$\begin{bmatrix} W'W & W'Z \\ Z'W & Z'Z + H^{-1}\frac{\sigma_e^2}{\sigma_a^2} \end{bmatrix} \begin{bmatrix} \hat{b} \\ \hat{a} \end{bmatrix} = \begin{bmatrix} W'y \\ Z'y \end{bmatrix}$$

Matrix H includes both the relationships, based on pedigree (A) and differences (A_δ) between those, and the genomic relationships, such that $H = A + A_\delta$. Thus, H is given by

$$H = \begin{bmatrix} A_{11} & A_{12} \\ A_{21} & G \end{bmatrix} = A + \begin{bmatrix} 0 & 0 \\ 0 & G - A_{22} \end{bmatrix},$$

where the subscripts 1 and 2 represent nongenotyped and genotyped individuals, respectively.

The inverse of H, which permits simpler calculations, is given by:

$$H^{-1} = A^{-1} + \begin{bmatrix} 0 & 0 \\ 0 & G^{-1} - A_{22}^{-1} \end{bmatrix} = \begin{bmatrix} A^{11} & A^{12} \\ A^{21} & G^{-1} + A^{22} - A_{22}^{-1} \end{bmatrix},$$

where A_{22}^{-1} is the inverse of the kinship matrix based on pedigree for only the genotyped individuals.

The estimator of the total genomic value for individual j is given by $\hat{a}_j = \sum_i X_{ij} \hat{m}_i$. This value, estimated when individual j does not participate in the estimation of m, can be correlated with the observed phenotype of j, seeking to conduct a validation. From the estimation of genetic values ($\hat{a}$) by GBLUP, the estimated marker effects ($\hat{m}$) can be obtained as described below:

$$\hat{a} = X\hat{m}$$
$$X'\hat{a} = X'X\hat{m}$$
$$\hat{m} = (X'X)^{-1}X'\hat{a}.$$

Models with dominance effects (d) can also be fitted. These are of the form $y = Wb + Xm + Td + e$. In this case, the elements of X are encoded as $(2)^{1/2}$, 0, and $-(2)^{1/2}$ for the genotypes AA, Aa, and aa, respectively. The elements of T are encoded as -1, 1, and -1 for the genotypes AA, Aa, and aa, respectively. Values of X and T encoded in this manner are independent and have a mean of 0 and a variance of 1. If the elements of X are encoded with the values -1, 0, and 1, the models with dominance effects will show the elements of T given by -0.5, 0.5, and -0.5 for the genotypes AA, Aa, and aa, respectively.

Analysis by GBLUP is computationally favorable because it results in fewer equations to be solved. Another important application of this analysis is the estimation of total heritability explained by all markers simultaneously. With the kinship matrix given by $G = (XX')/k = (XX')/[2\sum_i^n p_i(1 - p_i)]$, this h^2 can be estimated by REML using the mixed-model equations to estimate the variance components σ_a^2 and σ_e^2. The elements of matrix G represent the average multilocus kinship and are given by

$$G_{jk} = (1/n) \sum_{i=1}^{n} \frac{(x_{ij} - 2p_i)(x_{ik} - 2p_i)}{2p_i(1 - p_i)}$$

Another favorable feature of GBLUP is the possibility of directly estimating (by PEV) the accuracy of GWS. For individuals with phenotypes, this accuracy is valid for the estimation population without cross validation. In GBLUP, the validation population has its phenotypes replaced by missing data. Therefore, individuals from this validation population will have a validated accuracy estimate.

Models at the level of individuals including genotype $\times$ environment (ae) interactions can also be fitted, as long as there are related individuals within the same environment and also across environments. In this case, the model is equal to $y = Wb + Za + Zae + e$, where ae is the vector of effects from the interaction between additive genetic effects and environmental effects (random) and Z is the incidence matrix for a and ae. The mixed-model equations for predicting a and ae by the BLUP method are:

$$\begin{bmatrix} W'W & W'Z & W'Z \\ Z'W & Z'Z + G^{-1}\frac{\sigma_e^2}{\sigma_a^2} & Z'Z \\ Z'W & Z'Z & Z'Z + G_{ae}^{-1}\frac{\sigma_e^2}{\sigma_{ae}^2} \end{bmatrix} \begin{bmatrix} \hat{b} \\ \hat{a} \\ \hat{ae} \end{bmatrix} = \begin{bmatrix} W'y \\ Z'y \\ Z'y \end{bmatrix},$$

where $G_{ae} = G$ for pairs of individuals in the same environment and $G_{ae} = 0$ for pairs of individuals in different environments. The variance of the interaction between additive genetic effects and environmental effects is represented by σ_{ae}^2.

The GBLUP or genomic BLUP method can also be applied to address the heterogeneity of variance between markers. In this case, the matrix G is given by $G = (X^*DX^{*\prime})/\sigma_g^2$, where p_i is the frequency of one of the alleles of locus i and X^* is the matrix X corrected for the mean at each locus $(2p_i)$. The matrix D is given by $diag(D) = (\tau_1^2 ... \tau_n^2)$, and the elements τ_i^2 can be obtained using the methods IBLASSO, BLASSO, BayesA, BayesB, etc. This approach offers the following favorable features: (1) it permits simultaneous analysis of genotyped and nongenotyped individuals; (2) it allows direct calculation of the accuracy of selection by inverting the matrix of coefficients of the mixed-model equations; (3) the matrix D can be estimated in a single sample from the population and can be used for the entire selection population and in various generations; and (4) it allows dealing with the heterogeneity of genetic variance between markers.

INCREASE IN THE EFFICIENCY OF SELECTION FOR PLANT BREEDING

An increase in the efficiency of selection using GWS can be achieved by altering the four components of the expression for genetic progress, given by $G_S = (kr_{g\hat{g}}\sigma_g)/L$, where k is the standardized selection differential (dependent on the selection intensity), $r_{g\hat{g}}$ is the accuracy of selection, σ_g is the genetic standard deviation (genetic variability) of the trait in the population, and L is the time required to complete a selection cycle.

Perennial Plant Species and Species with Vegetative Propagation

In these species, the benefit of GWS is due to an increase in $r_{g\hat{g}}$ and a reduction in L. The increase in $r_{g\hat{g}}$ is due to the use of the actual kinship matrix proper to each trait (Resende, 2007). This increase depends on the size of the estimation population and marker density. The factor L is greatly reduced by GWS because the genomic prediction and selection can be performed at the seedling stage. Thus, even if $r_{g\hat{g}}$ is of the same magnitude as obtained by phenotypic selection, GWS is still better than selection based on phenotypes due to the reduction in L.

GWS in perennial species was applied by Valle et al. (2013), Resende Júnior et al. (2012a,b), and Resende et al. (2012) in forest species, by Cavalcanti and Resende (2012) in cashew trees, by Oliveira et al. (2012) in cassava, and by Simeão et al. (2013) in forage plants. Recent discussions on the methods applied are presented by De los Campos et al. (2012), Endelman and Jannink (2012), and Ober et al. (2012).

Allogamous Annual Plant Species

In allogamous annuals plants, the benefit of GWS is due to three factors: an increase in $r_{g\hat{g}}$, an increase in k, and a reduction in L. There is also an increase in the genetic variation exploited by the method of recurrent selection, as described by Fritsche-Neto (2011) and Fritsche-Neto et al. (2012).

In this case, the increase in $r_{g\hat{g}}$ results from using the actual kinship matrix and from the fact that all of the genetic variation in the population is exploited, not just the variation between families. If selection by GWS is performed early and before flowering, selection at the level of individuals and in both sexes becomes possible (as is performed for breeding perennial plants), without the need for two growing seasons: one to evaluate families and another to establish the recombination plot. Consequently, the time L is also reduced. This coincidence between the selection unit and the recombination unit maximizes the heritability of the selection method (exploits an additional 0.50 or 0.75 of the additive genetic variation that was within families). Selection at the individual level also yields an increase in the selection intensity k.

Autogamous Annual Plant Species

In these species, when haploid duplication is used to obtain inbred lines, the benefit of GWS is due to four factors: increase in $r_{g\hat{g}}$, increase in k, increase in σ_g (by exploiting twice as much additive genetic variation), and reduction in L.

According to the normal or genealogical method of breeding, it is accepted that selection by GWS cannot be performed on generation F_2 because breeding should continue until homozygosis for final selection. Thus, L is not reduced. However, the good alleles can be identified by GWS in generation F_2 to guide the crossing between the best plants, thereby performing intrapopulational recurrent selection in autogamous plants. This approach allows increasing $r_{g\hat{g}}$ and

σ_g, and genetic gain consequently increases. Additionally, k increases because it is possible to assess a much greater number of F_2 plants than $F_{2:3}$ families. To advance S_0 plants to inbred lines, early selection via GWS may be performed in each generation (without the need to experience the progeny in field trials), thereby maximizing the accuracy of selection. The estimation of marker effects are based on S_0 plants from generation F_2.

REFERENCES

Azevedo, C.F., Resende, M.D.V., Silva, F.F., Lopes, P.S., Guimaraes, S.E.F., 2013. Regressão via Componentes Independentes (ICR) para redução de dimensionalidade na seleção genômica para características de carcaça em suínos. Pesquisa Agropecuária Brasileira 48, 619–626.

Bernardo, R., Yu, J., 2007. Prospects for genome wide selection for quantitative traits in maize. Crop Science 47, 1082–1090.

Brondum, R.F., Su, G., Lund, M.S., Bowman, P.J., Goddard, M.E., Hayes, B.J., 2012. Genome position specific priors for genomic prediction. BMC Genomics 13, 543.

Cavalcanti, J.J.V., Resende, M.D.V., 2012. Predição simultânea dos efeitos de marcadores moleculares e seleção genômica ampla em cajueiro. Revista Brasileira de Fruticultura 34, 840–846.

Daetwyler, H.D., Villanueva, B., Bijma, P., Woolliams, J.A., 2008. Accuracy of predicting the genetic risk of disease using a genome-wide approach. PLoS ONE 3, e3395.

De los Campos, G., Hickey, J.M., Pong-Wong, R., Daetwyler, H.D., Calus, M.P.L., 2012. Whole genome regression and prediction methods applied to plant and animal breeding. Genetics (available only ahead of print, June 28 doi:10.1534/genetics.112.143313).

Dekkers, J.C.M., 2004. Commercial application of marker and gene assisted selection in livestock: strategies and lessons. Journal of Animal Science 82, 313–328.

Dekkers, J.C.M., 2007. Prediction of response to marker assisted and genomic selection using selection index theory. Journal of Animal Breeding and Genetics 124, 331–341.

Endelman, J.B., Jannink, J.L., 2012. Shrinkage estimation of the realized relationship matrix. Genes, Genomes, Genetics 2, 1405–1413.

Erbe, M., Hayes, B.J., Matukumali, L.K., Goswami, S., Bowman, P.J., Reich, C.M., et al., 2012. Improving accuracy of genomic predictions within and between dairy cattle breeds with imputed high-density single nucleotide polymorphism panels. Journal of Dairy Science 95, 4114–4129.

Fernando, R.L., Grossman, M., 1989. Marker-assisted selection using best linear unbiased prediction. Genetics Selection Evolution 21, 467–477.

Fritsche-Neto, R., 2011. Seleção genômica ampla e novos métodos de melhoramento do milho. UFV, Viçosa. 28p. (Tese de Doutorado).

Fritsche-Neto, R., Dovale, J.C., Resende, M.D.V., 2012. Genome wide selection for root traits in topical maize under stress conditions of nitrogen and phosphorus. Acta Scientiarum Agronomy 34 (4), 389–395.

Garrick, D.J., Taylor, J.F., Fernando, R.L., 2009. Deregressing estimated breeding values and weighting information for genomic regression analyses. Genetics Selection Evolution 41, 55.

Gianola, D., de los Campos, G., 2009. Inferring genetic values for quantitative traits non-parametrically. Genetics Research 90, 525–540.

Gianola, D., Perez-Enciso, M., Toro, M.A., 2003. On marker-assisted prediction of genetic value: beyond the ridge. Genetics 163, 347–365.

Gianola, D., Fernando, R.L., Stella, A., 2006. Genomic-assisted prediction of genetic value with semiparametric procedures. Genetics 173, 1761–1776.

Gianola, D., Campos, G., Hill, W.G., Manfredi, E., Fernando, R., 2009. Additive genetic variability and the Bayesian alphabet. Genetics 183, 347–363.

Goddard, M.E., Hayes, B.J., 2007. Genomic selection. Journal of Animal Breeding and Genetics 124, 323–330.

Grattapaglia, D., Resende, M.D.V., 2010. Genomic selection in forest tree breeding. Tree Genetics and Genomes http://dx.doi.org/10.1007/s11295-010-0328-4.

Guo, Z., Tucker, D.M., Lu, J., Kishore, V., Gay, G., 2012. Evaluation of genome-wide selection efficiency in maize nested association mapping populations. Theoretical and Applied Genetics 124, 261–275.

Lande, R., Thompson, R., 1990. Efficiency of marker-assisted selection in the improvement of quantitative traits. Genetics 124, 743–756.

Legarra, A., Robert-Granié, C., Croiseau, P., Guillaume, F., Fritz, S., 2011. Improved Lasso for genomic selection. Genetics Research 93 (1), 77–87. Cambridge.

Meuwissen, T.H.E., 2007. Genomic selection: marker assisted selection on genome-wide scale. Journal of Animal Breeding and Genetics 124, 321–322.

Meuwissen, T.H.E., Hayes, B.J., Goddard, M.E., 2001. Prediction of total genetic value using genome-wide dense marker maps. Genetics 157, 1819–1829.

Misztal, I., Legarra, A., Aguilar, I., 2009. Computing procedures for genetic evaluation including phenotypic, full pedigree, and genomic information. Journal of Dairy Science 92 (9), 4648–4655.

Moser, G., Khatkar, M., Hayes, B., Raadsma, H., 2010. Accuracy of direct genomic values in Holstein bulls and cows using subsets of SNP markers. Genetics Selection Evolution 42, 37.

Ober, U., Ayroles, J.F., Stone, E.A., Richards, S., Zhu, D., et al., 2012. Using whole-genome sequence data to predict quantitative trait phenotypes in *Drosophila melanogaster*. PLoS Genetics 8 (5), e1002685. http://dx.doi.org/10.1371/journal.pgen.1002685.

Oliveira, E.J., Resende, M.D.V., Santos, V.S., et al., 2012. Genome-wide selection in cassava. Euphytica 187, 263–276.

Resende, M.D.V., 2007. Matemática e estatística na análise de experimentos e no melhoramento genético. Embrapa Florestas, Colombo. 435p.

Resende, M.D.V., 2008. Genômica quantitativa e seleção no melhoramento de plantas perenes e animais. Embrapa Florestas, Colombo. 330p.

de Resende, M.D.V., Lopes, P.S., Silva, R.L., Pires, I.E., 2008. Seleção genômica ampla (GWS) e maximização da eficiência do melhoramento genético. Pesquisa Florestal Brasileira 56, 63–78.

Resende, M.D.V., Resende Júnior, M.F.R., Aguiar, A.M., Abad, J.I.M., Missiaggia, A.A., Sansaloni, C., et al., 2010. Computação da seleção genômica ampla (GWS). Embrapa Florestas, Colombo. 79p.

Resende, M.D.V., Silva, F.F., Viana, J.M.S., Peternelli, L.A., Resende Júnior, M.F.R., Valle, P.R.M., 2011. Métodos estatísticos na seleção genômica ampla. Embrapa Florestas, Colombo. 106p.

Resende, M.D.V., Resende Junior, M.F.R., Sansaloni, C., Petroli, C., Missiaggia, A.A., Aguiar, A.M., et al., 2012. Genomic selection for growth and wood quality in eucalyptus: capturing the missing heritability and accelerating breeding for complex traits in forest trees. New Phytologist 194, 116–128.

Resende Júnior, M.F.R., Valle, P.R.M., Resende, M.D.V., Garrick, D.J., Fernando, R.L., Davis, J.M., et al., 2012a. Accuracy of genomic selection methods in a standard dataset of loblolly pine. Genetics 190, 1503–1510.

Resende Júnior, M.F.R., Valle, P.R.M., Acosta, J.J., Peter, G.F., Davis, J.M., Grattapaglia, D., et al., 2012b. Accelerating the domestication of trees using genomic selection: accuracy of prediction models across ages and environments. New Phytologist 193, 617–624.

Simeão, R.M., Casler, M.D., Resende, M.D.V., 2013. Genomic selection in forage breeding: designing an estimation population. In: Plant and Animal Genome Conference XXI – 2013, San Diego.

Solberg, T.R., Sonesson, A., Wooliams, J., Meuwissen, T.H.E., 2006. Genomic selection using different marker types and density. In: Proceedings 8th World Congress of Genetics Applied to Livestock Production, Belo Horizonte, Ed. da UFMG, 1 CD-ROM.

Sved, J.A., 1971. Linkage disequilibrium and homozygosity of chromosome segments in finite populations. Theoretical Population Biology 2, 125–141.

Valle, P.R.M., Resende Júnior, M.F.R., Huber, D.A., Quesada, T., Resende, M.D.V., Neale, D.B., et al., 2013. Genomic relationship matrix for correcting pedigree errors in breeding populations: impact on genetic parameters and genomic selection accuracy. Crop Science 53.

Van Raden, P.M., 2008. Efficient methods to compute genomic predictions. Journal of Dairy Science 91 (11), 4414–4423.

Vazquez, A.I., Rosa, G.J., Weigel, K.A., de los Campos, G., Gianola, D., Allison, D.B., 2010. Predictive ability of subsets of SNP with and of parent average for several traits in US Holsteins. Journal of Dairy Science 93 (1), 5942–5949. http://dx.doi.org/10.3168/jds.2010-3335.

Whittaker, J.C., Thompson, R., Denham, M.C., 2000. Marker assisted selection using ridge regression. Genetical Research 75, 249–252.

Chapter 6

Genes Prospection

Valdir Diola[a] and Roberto Fritsche-Neto[b]

[a]Federal Rural University of Rio de Janeiro, Department of Genetics, Seropédica, Rio de Janeiro, Brazil, [b]University of São Paulo (USP), Department of Genetics, São Paulo, Brazil

INTRODUCTION

Until 1960, mathematical and statistical analysis of phenotypic characteristics was the basis for genetic studies because it allowed the prediction and monitoring of inheritable characters transmission. Genetic maps derived from morphological markers were one of the main strategies to estimate these parameters since they provided connection and position information in genome. They represented a high cost, since they depended on extensive observation periods, several cycles, large populations, and high inaccuracy rate in phenotyping.

From cytogenetic observations and recombination rate analyses, it was determined that genetic factors were limited to smaller regions on the chromosome, the gene loci. A eukaryote gene comprising the region in a DNA strand that comprises a promoter sequence (responsible for the assembly of the RNA polymerase complex) and the transcribed region can reproduce a phenotype. The phenotypic expression is the result of the hereditary material assembly and the expression control of corresponding genes. With the techniques of molecular markers, transmission monitoring of hereditary traits was made easier. It became possible to identify DNA sequences correlated with the phenotype and to position them in linkage groups.

Initially, extensive genetic maps needed to be built in order to detect and to get closer to these loci, which in most cases were located a few hundred kilobases (kb) from the nearest markers. Currently, a number of molecular techniques are suitable for the identification of regions of interest and they offer a wide range of genetic information that facilitate a more specific and faster approach, reducing time and labor.

Despite the considerable progress, gene prospecting is a great challenge, because the accurate characterization of phenotypes and genotypes is critical to the success of the works to be achieved. Gene prospecting requires from the professional or from the team extensive knowledge in various areas, including integrating classical plant breeding, molecular markers, biochemistry, plant physiology, biotechnology, and bioinformatics (Figure 6.1). The final result of

Biotechnology Applied to Plant Breeding. http://dx.doi.org/10.1016/B978-0-12-418672-9.00006-4

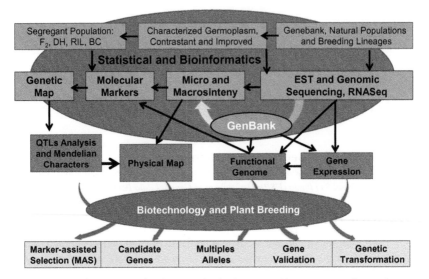

FIGURE 6.1 Integration of molecular strategies for the genes of interest prospections under a set of techniques and applications in the study of characterization and isolation of gene loci and their application in plant breeding and biotechnology.

the biological model, the ideotype plant (desired phenotype), will be a plant breeding point of interest.

PRINCIPLES OF GENETIC MAPPING

The isolation of the genes starts with the existence of polymorphism for phenotypic character to get to the isolation of specific loci. The technique is based on the recombination rate, in others words, the probability of occurrence crossing-overs at meiosis metaphase I during gamete formation in parents. Chromosomal transfer promoted by a series of recombinases enzymes may occur during meiosis metaphase I. The effect of balance recombination (occurring at the same locus in both sister chromatids) is only noticeable if the recombinant locus is heterozygous, because the alleles are different and create new allele combinations. The meiotic crossing-overs have the effect of increasing the genetic variability and form new allelic combinations.

We will explain the events and probability of genetic recombination using diploid lines of tomato, and a breeding program that used two dominant characteristics in contrasting parents. Progenitor 1 is susceptible to fungal early blight (alleles pp) but has no incidence of apical fruit necrosis (blossom rot tolerant) (without Calcium deficiency: CC alleles) (genotype ppCC); progenitor 2 is early blight resistant and has high blossom rot incidence in the fruit (genotype PPcc). In scheme crossing the following segregation was detected (Table 6.1).

TABLE 6.1 Phenotypic Segregation for Characteristics of the Resistance to *Alternaria solani* (Early Blight) and Tolerance to Tomato Calcium Deficiency (Blossom End Rot)

Phenotypes	Genotypes	Generations				
		P_1	P_2	F_1	F_2	Test Crossing
Resistant and Tolerant –Ca	P_C_			32	254	38
Resistant and Sensible –Ca	P_cc		5		118	200
Susceptible and Tolerant –Ca	ppC_	5			122	250
Susceptible and Sensible –Ca	Ppcc				6	12
Total		5	5	32	500	500

In the F_1 generation, recombinant individuals were not observed, because the parents are homozygous for alleles (PPcc) and (ppCC), so even if there was recombination during gametogenesis, recombinant gametes are similar. In the F_2 population, you can see that there are recombinant genotypes (ppcc), because there would be no swaps cp gametes in F_1. Thus, it is concluded that the gametes cp (both recessive alleles on the same chromosome) are recombinants and genotype ppcc is the product of two recombinant gametes. The recombination rate is evaluated by a crossing test where this double-recombinant (ppcc) is crossed with genotypes of the population F_1 generation (PpCc), as the rate of recombination of gametes F_1 genotype is evaluated (possible to be detected) (Figure 6.2).

Then the recombinant phenotypes observed for the test cross are 38 resistant to early blight and tolerant to calcium deficiency (PpCc) and 12 double recessive (ppcc), totaling 50 recombinant genotypes in 500. Therefore, the recombination rate of the gametes in progenitors F_1 = number of recombinant test crossing/genotypes; Rec. $= (50/500) * 100 = (0.1) * 100 = 10\%$ of gametes recombinant, probably 5% gametes PC and 5% pc (if one chromatid occurs, crossing-over affects the sister chromatid) and 45% Pc and 45% PC parental (non recombinant). Thus the rate of recombination in the region comprising the gene for early blight resistance and the gene coding for tolerance to calcium deficiency is 10%.

Test the segregation of the marker with the chi-square ($\chi^2 = (Fo-Fe) 2/Fe$), where Fe is the frequency expected and Fo is the frequency observed. Being a monogenic character, both genes in the F_2 generation have $254 + 118 = 372$ genotypes resistant to early blight and $122 + 6 = 128$ susceptible; then

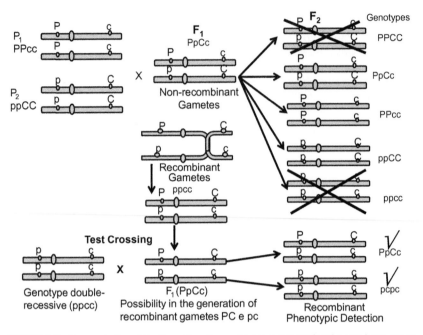

FIGURE 6.2 Demonstration of the formation of recombinant gametes using the crossing test; non recombinant genotypes: ppCc and Ppcc (not shown); and recombinant: PpCc and ppcc.

$372/128 = 2.90$, so the ratio is $\sim 3{:}1$ (four types of possible gametes PP (25%), Pp (2x)(50%), and pp (25%)) and the expected frequency of 125 individuals would be for each, and for the resistance would be $3 \times 125 = 375$ genotypes P_ (resistant) and 125 pp (susceptible). Thus, for resistance:

$$\chi^2 = (Fo - Fe)\,2/Fe = (372 - 375)\,2/375 = (3)\,2/375 = 9/375 = 0.024$$

and the susceptibility of the χ^2 is 0.072. Therefore, the sum of both is 0.096. Considering two phenotypic classes, resistance and susceptibility to early blight, the freedom degree is 1 (one class can take any random value, but the other class will have the value determined in order to complete the total of individuals). Thus it refers to a probability table for χ^2 (Table 6.2).

Thus, with one degree of freedom the calculated value of χ^2 is at 0.096, and is between 70 and 80% probability that the random deviations, thus the observed segregation in the F_2 generation, adjusts the ratio of 3:1 and therefore the resistance trait, early blight, is controlled by a dominant gene. For tolerance calcium deficiency, the calculated $\chi^2 = 0.010$, is similar to previous analysis.

Genetic mapping is dependent on at least three markers and phenotypic traits on the same linkage group, to enable the ordering of markers on chromosome, corrections for segregation distortion, and the detection of double recombinants

TABLE 6.2 Probability Level for χ^2 Test

GL	Probability										
	0.95	0.90	0.80	0.70	0.50	0.30	0.20	0.10	0.05	0.01	0.001
1	0.004	0.02	0.06	0.15	0.46	1.07	1.64	2.71	3.84	6.64	10.83
2	0.10	0.21	0.45	0.71	1.39	2.41	3.22	4.60	5.99	9.21	13.83
3	0.35	0.58	1.01	1.42	2.37	3.66	4.64	6.26	7.28	11.34	16.27
	Non significant								Significant		

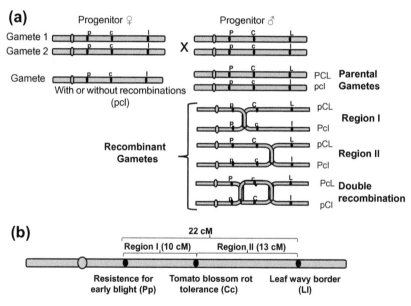

FIGURE 6.3 Structuring the study of genetic mapping. a) Test Crossing: the gametes completely recessive genotype does not allow the observation of recombination, because not generates allelic variability, with all gametes pcl. The detection of recombination is only possible genotypes in the gametes derived from the F_1 generation, requiring three points for observing the double recombinants. b) Genetic mapping for spatial positioning of loci according to the sort method of the Sun Adjacent Recombinant Fraction (SARF) using three points.

(Figure 6.3 and Table 6.3). Figure 6.3 represents a linking group with three loci positioned and the respective distances between them. The distances do not add up to 1%; recombination is not necessarily 1 cM (centimorgan). That is because the points are based on calculations of adjusted positioning of loci first two to two and then to the nearest neighbor based on SARF (Sum Adjacent Fraction Recombinant), accommodating them on an imaginary line. Referring to the

TABLE 6.3 Phenotypes Observed in the Descendants of the Test Crossing

Phenotypes in the Progeny	Gamete ♀	Gamete ♂	Freq. Obs.
Normal	Pcl	PCL	200
Sensible –Ca and Leaf wavy border	Pcl	Pcl	24
Leaf wavy border	Pcl	PCl	32
Susceptible early blight and Leaf wavy border	Pcl	pCl	2
Susceptible early blight, Sensible –Ca and Leaf wavy border	Pcl	pcl	190
Sensible –Ca	Pcl	PcL	3
Susceptible early blight and sensible –Ca	Pcl	pcL	28
Susceptible early blight	Pcl	pCL	21
Total	---	---	500

previous example, add the morphological leaf border, smooth to dominant (L_) and wavy to recessive (ll). In this example, we will do the calculations with 500 plants obtained from crossing test (genotype PpCcLl × ppccll) (Figure 6.3). The male gametes containing alleles PCL or pcl and female gametes are completely recessive pcl (Table 6.3 and Figure 6.3).

Paternal combinations	PCL = 200	e	pcl = 190
Double recombinations	PcL = 3	e	pCl = 2
Recombinations in the Region I	pCL = 21	e	Pcl = 24
Recombinations in the Region II	pcL = 28	e	PCl = 32

Region I (%) = (Region I + double)/total = (21+24+3+2)/500 = 0,1 × 100 = 10%

Region II (%) = (Region II + double)/total = (28+32+3+2)/500 = 0,13 × 100 = 13%

Although these observations allow the detection of double recombinants, often by low accuracy phenotyping or by insufficient individuals in the mapping population, some doubles recombination is not perceived. Calculate the inference to adjust the real values of occurrence of double recombination. It is based on the following calculation:

Frequency of double recombinations observed:
FPDO = (No double recombinants/total progeny n) = $100 \times [(2+3)/500] \times 100 = 1\%$
Expected frequency double recombinations
FPDE = (dist. region I (%) $\times$ dist. Region II (%)) $\times 100 = (10 \times 13) \times 100 = 1.3\%$
Estimation of inference (i)
i = $[1 - (FPDO / FPDE)] \times 100 = [1 - (1/1.3)] \times 100 = 0.23 \times 100 = \sim 23\%$ of the frequency of double recombinants were not observed.

The estimation of double recombination observed total is 1% over 23% of this value (i.e., 1.23% of double recombination) and because of this it makes adjustments, typically using the statistical method of Bonfferoni.

The Phenotyping of a Population

The phenotype can be understood as the result of the direct effect of the interaction between two sources of variation: the genotype and the environment. Until now, plant breeding has obtained more significant results considering genetic aspects based on phenotype knowledge, which helps to detect traits of interest. Accurate phenotyping is one of the most critical phases, since many errors may be accumulated and methods of phenotyping in most cases are based on visual analysis. Some improvement programs already use biochemical markers linked to genes of interest, or even to phenotyping at a phenomic large scale, which is more accurate. However, little automation is available, since the variation in analysis is very wide. Currently and in the future, with the development of molecular genetics tools, phenotyping may be one of the bottlenecks of breeding programs.

There are three important points in phenotyping for most studies: (1) phenotypic parameter (discriminating trait), (2) conditioning environmental effect, and (3) genetic background and developmental stage of the species assessed.

Mapping Populations

The first step in producing a mapping population is the selection of two genetically different parents who show clear genetic differences (phenotypic and molecular) to one or more characteristics of interest. It is important that parents do not exhibit gametic incompatibility or preferential matting, nor show low fertility, lethal genes, or mortality, because this will affect the analysis of segregation and therefore, the accuracy of gene or QTL identification. Progenies of

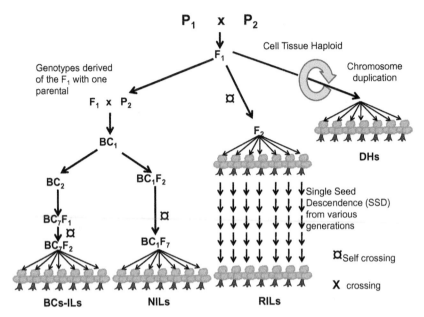

FIGURE 6.4   Diagram of the main methods for obtaining genetic mapping populations. F_2 populations are developed by self-crossing F_1 hybrids, BC is produced by crossing the F_1 progeny with a parent (or the receiver recurrent parent). RIL are developed by the method of driving populations SSD (SSD-Single Seed Descendent). BCnFn, is the result of sequential backcrossing followed by self-pollination, from which arise the Near Isogenic Lines (NIL). The double-haploid (DH) is the result of monoploid genotypes derived from genomic duplication. (*Adapted from Semagn et al., 2006*)

the second generation (F_2), backcrossing (RC), recombinant inbred lines (RIL) and double haploid (DH), and near isogenic lines (NIL), are most frequently used for genetic mapping (Figure 6.4) (Semagn et al. 2006).

Populations derived from the same parents at different times may have different recombination rates, which is slightly affected by changes in the environment. In the most critical moments, genetic recombination is higher as a result of the increased activity of recombinases, seeking to increase genetic variability as an evolutionary runaway process that affects phenotype. In this regard, He et al. (2001) compared linkage maps between DH and RIL populations derived from the same crossing in rice, but obtained in different environmental conditions. The size of the map in RIL population was lower than the one in DH population (70.5%).

F_2 and RC are the simplest types of mapping populations because they are easy to build and require only a short period of time to be obtained. However, they are highly heterozygous and cannot be indefinitely propagated by seeds. RIL, NIL, and DH are permanent populations, they are homozygote lineages that can be multiplied and reproduced with minimal occurrence of genetic change, however, they do not allow us to estimate the genetic component of dominance deviations. Moreover, they may show deviations on segregation proportion due to the natural selection on the plants during the long period to obtain mapping population.

The optimal number of individuals in the population is indeterminate. Simulation studies considering sample sizes ranging from 50 to 1000 individuals of F_2, RC, RIL, and DH populations indicate that the type and size of mapping populations can exert significant influence on the accuracy of genetic maps (Ferreira et al., 2006):

I. Populations with a low number of individuals result in a large number of fragmented linking groups and the determination of loci is inaccurate.
II. More accurate maps are obtained for RIL and F_2 population with co-dominant markers. Maps constructed from F_2 with dominant markers are less accurate;
III. The greater the number of individuals or markers, the more accurate the map because saturation better adjusts loci in their positions.
IV. For all types of populations, a minimum number of 200 individuals are required to build reasonably accurate linkage maps.
V. If the population is small for the intended goal or the genetic variability is high, it is necessary to increase substantially the number of markers or genotypes.

In practice, the population size used in preliminary genetic mapping studies ranges from 50 to 250 individuals, but when the goal is the exploration of alleles a larger population is needed for high-density mapping (Tanksley et al., 1995). It is important to note that, in order to achieve greater accuracy in the analysis it is necessary to balance the number of markers and the population size. So, there is no use in having a few individuals and many prints, or even many individuals and few prints. Note then that it is necessary that the number of parameters to be estimated (markers) have values consistent with the number of observations (progeny size).

In general, maximum amount of genetic information is obtained from F_2 populations when a system of co-dominant markers is used. This is because this type of population shows the maximum linkage disequilibrium and co-dominant markers allow us to estimate the effect of dominance deviations with greater accuracy. As for dominant markers, they provide similar information to co-dominant ones in RIL, NIL, and DH because all loci tend to be homozygous (Semagn et al., 2006).

ISOLATION OF GENES BY USING TECHNIQUES FOR STRUCTURAL GENOME

All kinds of markers can be used in gene mapping in plants. The first molecular marker used was the RFLP and its main advantage is to consider co-dominant inheritance, a specificity of loco and easily reproducible. With the development of polymerase chain reaction (PCR), genetic mapping was facilitated. Microsatellite markers (SSR) are now used on a large scale, since they have advantages such as high informativeness, co-dominant inheritance, reproducibility and ease of automation for high-performance. AFLP markers are widely used because they are highly polymorphic, reproducible, and they can be used for any organism without a high initial investment in primers. They generally provide good genomic coverage, however they are insensitive to DNA methylation, hindering analyzes in the region of heterochromatin. RAPD markers are not widely accepted because their reproducibility is questionable. AFLP, RAPD, and ISSR share a common limitation for linkage mapping, because they depend on a secondary marker for their application to genetic improvement. They are dominant markers, unable to show the differences between dominant homozygous and heterozygous. These markers are less valuable for mapping F_2 populations or RC. DART is a fast, highly polymorphic and reproducible high-performance method, but its dominant inheritance is still a limitation for mapping. SNPs are numerous, quite polymorphic; when in haplotype level (1 SNP in every 500-1000 bp) it is highly transferable and of co-dominant inheritance. SNP genotyping technologies on a large scale have become frequently used in recent years. As a result, a wide variety of different genotyping protocols based on SNP was made available. There is not a single protocol that meets all the needs of different aspects of investigation in determining the best SNP technology regarding sensitivity, reproducibility, accuracy, multiplexing capability for the analysis of high-performance, and the cost-effectiveness relation in an initial investment on equipment.

Other alternatives for reducing mapping work and maximizing results shortening the time for obtaining the maps are based on the use of BSA (Bulked Segregant Analysis) technique (Michelmore et al, 1991). It is a technique that uses the pool of DNA from a genotype set of similar phenotypic characteristics in the same PCR reaction, comparing the parents and two segregating bulks for the gene or character in study (Figure 6.5a). In the study of a gene mapping for coffee rust resistance, Diola et al. (2011) used BSA technique in a population of 224 genotypes in the F2 generation through AFLP markers. Genotyping with closed bulks consisted of the analysis of 1,154 combinations of AFLP selective primers, allowing the analysis of approximately 80,000 loci. Of these, 363 showed parent polymorphism and resistant bulk. After analyzing the 104 polymorphic loci through bulks opening, only 25 markers showed the expected polymorphism (Figure 6.5) and they were applied on the population; they were all connected (Figure 6.6).

Linkage Analysis and Map Construction

These stages are performed in computer programs, since data processing is relative to the number of markers and genotypes considered. The most widely used programs are JoinMap (http://www.kyazma.nl/index.php/mc.JoinMap), MapMaker/EXP(http://linkage.rockefeller.edu/soft/mapmaker/),GMENDEL (http://www.bio.net/mm/gen-link/1994-March/000308.html), LINKAGE

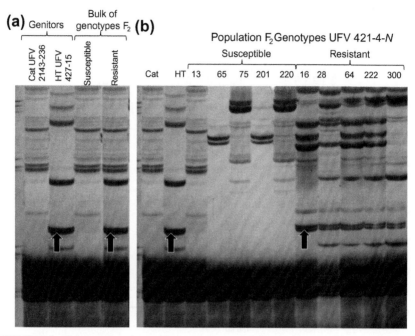

FIGURE 6.5 Polymorphism detected between genotypes resistant and susceptible to Coffee Leaf Rust using a combination of AFLP primer ETCA / EATG. (a) Amplicon only in the resistant parent and resistant bulk. (b) Fragment present in the progenitor and all genotypes of the bulk resistant and absent in the bulk susceptible. (*Adapted from Diola, 2011*)

(ftp://linkage.rockefeller.edu/software/linkage), and MapDisto (http://mapdisto. free.fr/). JoinMap is a commercial program, while all others are available for free. The basic principles of map building are basically the same for the different statistical programs and the important steps for linkage analysis are described as follows:

Test for segregation distortion: For each segregating marker, a chi-squared analysis must be performed in order to test segregation deviation (1:1 to dominant and co-dominant markers in BC, RIL, DH, and NIL; 1:2:1 to co-dominant markers or 3:1 for dominant in F_2). A significant deviation from the expected genotype frequencies is called segregation distortion, which can occur by statistical bias; technical and genotyping errors; or biological issues such as chromosome loss, competition among gametes, preferential fertilization, gametocyte genes or pollen's killer (male or female gametes abortion), incompatibility genes, and chromosome arrangements by non homologous pairing. The effects of adding loci with distortion in the genetic map have little effect both on the marker's order and on the map size. DH and RIL populations have high segregation distortions regarding BC. In DH, they occur due to selection pressure associated with the *in vitro* production process and recessive lethal genes fixation in homozygosis.

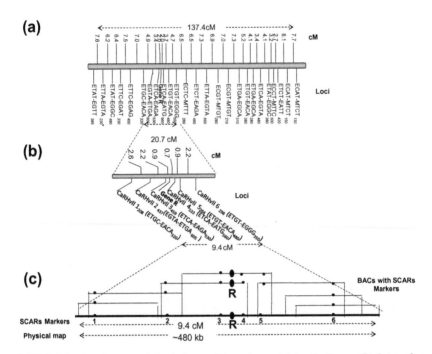

FIGURE 6.6 Genetic map and physical map of the region containing the R gene (R) that confers resistance of coffee plant to race II of *H. vastratrix* (Coffee Leaf Rust). (a) Genetic map saturated with AFLP markers, χ2 * Significant segregation of 3:1. (b) High-density genetic map with SCAR markers. (Plotting the format "Mapmaker" the MapDisto) (Diola et al., 2011). (c) Physical map obtained sorting the contigs of BACs.

Establishment of linkage groups: Markers are indexed to linkage groups by using LOD ratio, which refers to the logarithm of the probability that two loci are in the same group at a given ratio and that they are not connected (LOD score). Pairs of markers with a LOD recombination score above a critical linklod are considered associates. On the other hand, those with LOD scores lower than the critical one are considered of independent segregation. Several researchers use it as critical value 3 (i.e., as the minimum limiting value in order to consider the non independent segregation of a marker regarding a linkage group). A LOD value of 3 between two markers indicates that the probability of linkage between them is 1000 times greater than the probability that they segregate independently (Stam, 1993). In this regard, when we consider LOD high values, it will result in a greater number of fragmented linkage groups, each with fewer markers, whereas small values of LOD tend to create short linkage groups, with a large number of markers per group. In practice, the determination of linkage groups is generally not a simple task, because the level of rigor can cause distortion in the analysis. Two or more markers may be attached at more than one linkage group very common in polyploids, particularly in autopolyploids. In general, we seek that the number of map's linkage groups is equal to the haploid number of species chromosomes.

Determination of distance and order of loci in the map: In order to calculate the distances and the order of markers in linkage groups, it is necessary to specify several parameters, including a recombination limit value, a minimum LOD and mapping function. Only marker pairs information of with a LOD score above the critical values are used to calculate the distances. The choice of these values is arbitrary and it is advisable not to use values below 3.0 and a recombination rate lower than 30%. In the case of small populations or few markers, it is recommended to reduce rigor. Mapping procedure is basically a process of building a map by adding loci, one by one, from the most informative pair of loci (more tightly linked pair of loci). If the order of marker sets (at least three) is known in advance, this information can be provided to the program as a "fixed order." At each stage, a marker is added to the map based on its total linkage information with the markers that have been previously placed on the map. For each locus added, the best position is estimated and an adjustment likelihood measure is calculated. When the adjustment likelihood reduces significantly, or when the locus shows negative values (negative distances), they must be recalculated with new parameters. The alternative is to delete the locus and test it again at the end, after the first round, in which all loci previously removed are tested in order to be added to the map in a subsequent stage. Care should be taken so that these inclusions do not modify marker order, small changes in the distances occurring between the loci closer to the insertion. All loci removed are included in a final stage, with less accuracy in mapping parameters, discarding loci with negative distances.

When the distances between the markers on the map are small (<10 cM), this distance is equal to the recombination frequency. However, using mapping functions, Haldane or Kosambi, is recommended, which adjusts the values of recombination frequencies at map distances and vice versa. Haldane mapping function (1931) assumes the absence of crossing-over interference in meiosis, whereas Kosambi mapping function (1944) assumes a certain degree of interference (non detected double recombination). Mapping functions convert recombination fractions into map units, centimorgans (cM), or map unit (mu). Generally, Haldane function obtains distance estimates higher than those obtained by Kosambi, which, consequently, generate maps of greater length. The statistical programs perform the ordering of loci using weighted least squares, maximum likelihood and minimum sum of adjacent recombination fractions (SARF) criteria. The distance becomes greater when using the criterion of maximum likelihood regarding the others due to its rigor. The advantage of weighted least squares is that the distances between markers are calculated from the map distances between all pairs of markers on a chromosome and, therefore, the impact of errors on the distance between adjacent markers is less severe. SARF is recommended for high-density mapping, for ordering markers in pairs from the nearest neighbor.

RELATIONSHIP BETWEEN GENETIC AND PHYSICAL MAPS

Genetic maps represent genome structure according to the recombination rate in each specific locus, while physical maps represent the sequence order of nucleotides in the genome. Generally, linked molecular markers correlate with genetic distances and the physical placement of the loci in the same order with small variations between relative genotypes. Locos highly saturated or co-segregating markers remain stable in the physical map. However, a direct linear relationship between genetic distance units in centimorgan (cM) and physical distances in base pairs (bp) does not exist. Regarding chromosome 4 of Arabidopsis, for example, 1 cM varied from 30 to 550 kb (Schmidt et al. 1995). For the rice, a 1 cm measure represented in average the physical distance of 258.5 kb, ranging from 120 to 1000 kb per cM (Kurata et al. 1994). In their study, Diola et al. (2011) using SCAR markers derived from AFLP, the specific region of a coffee rust resistance locus showed that 1 cM was approximately 50 kb (Figure 6.6) to the chromosomal locus in question.

In this case, during gametogenesis, 1% of gametes showed recombination of 50 kb in this region and thus 9% showed recombination of 450 kb in the region. However, this measure does not allow us to extrapolate this statement to different populations (BC, RIL, NIL, DH, etc.), derived from different parents, or even if coming from the same parents but obtained in different environments. However, genetic mapping provides a rough idea of the recombination rate for a particular locus that may, with certain restrictions, be applied to other crossings in correlated breeding programs.

Normally, these distances are not fixed in the genome, because in the euchromatin regions (regions less dense) 1 cM represents a physical distance much lower than in heterochromatic regions (denser regions comprising centromeres and telomeres and also internal regions of the chromosome arms). This fact is especially due to the degree of DNA methylation and the difficulty of access of the recombinases to these regions during metaphase I of meiosis in gamete formation. The recombination rate is also affected by many other factors, including the environment; the more severe it is the higher the rate of recombination aiming to expand the set of possible allelic combinations.

PHYSICAL MAPPING AND ISOLATION OF LOCI OF INTEREST

It is not always necessary to know the sequence of locus of interest region; however this information can contribute to the development strategies of gene cloning, validation of genes of interest, identification of co-segregating markers, and specific SNP. The isolation of regions of interest consists in a genome reduction, in which all that matters is the identification of the specific spot sequence.

There are several isolation methodologies and positional cloning of gene(s). The most widespread ones depend on the use of genomic libraries of BAC (Bacterial Artificial Chromosome). The size of these libraries is determined by the number of clones related to the genome size, so that a good genome coverage is approximately five times its size (i.e., 450 Mb genome, 120 kb BAC; 3750 clones are necessary for represent the same sample content of the genomic DNA X 5 = 18,750 clones). BAC clones carry ideally genomic fragments of approximately 120 kb. Thus, for a genome coverage containing 1 Gb about 8000 BAC clones are necessary, multiplied by the number of times that genome must be covered (5x = 40,000 clones). Isolation of BAC clones is performed in different ways, hybridization or PCR being the most common. Through hybridization, membranes containing the DNA or BAC clones hybridized with specific probes are related to markers closely linked to the locus of interest. Using the BAC clones plates three-dimensional method by PCR facilitates the isolation of clones carrying the sequences of interest in several reductional stages. Initially, there are a large number of clones (e.g., 384 clones), followed only with the PCR that amplify target sequences (co-segregating markers) until reaching the specific clone.

Until mid-2006, BAC clones sequencing was a complex task because sequencers were capable of producing up to 800 bp sequences and clones contained $\sim$ approximately 120 kb. Techniques such as BAC end, BAC fingerprint, BAC Shotgun, Chromosome Walking, and Chromosome Landing were widely applied, and still remain well accepted. Each of them has a specific purpose. BAC end is used to detect overlaps on the ends of duplicate clones and clones. BAC Shotgun and BAC fingerprint are based on BAC clone fragmentation, sequencing, and alignment of sequences for contigs assembly (long DNA sequences obtained from overlapped fragments alignment). The two other techniques are based on BAC clone sequencing in several stages, always taking the end of the DNA as a reference (Chromosome Walking) or starting from markers that are closer to the locus of interest (Chromosome Landing).

Currently, with the advancement of sequencing technologies such as NGS (Next Generation Sequencing), the sequencing of large fragments such as BAC clones is possible, since they are able to generate millions of sequences. Pyrosequencing in 454 FLX Titanium units generates approximately 500,000 Reads up to 400 bp, sufficient to cover up to eight times a pool containing up to 200 BAC clones. On the other hand, devices such as Illumina HiSeq 2000 allows the fractioning of a plate into eight parts generating five million fragments containing from 75 to 100 bp (corresponding to 4 to 5 Gb = 41,000 clones of 120 kb), enough to cover 20 times a pool of about 2000 BAC clones. As it reduces the size of the amplicons sequenced, one must increase the necessary genomic coverage because of the requirements for alignment. However, short sequences are problematic when BAC clones contain long sequences of repetitive DNA.

In both the cases, using automatic sequencers or NGS bioinformatics programs are required for the alignment and construction of contigs. Many programs are used to generate the Genome Assembly (alignment) of the contigs (CCL workbenk, CAP3, etc.). Starting from the sequences the following phases are: identification of candidate genes, test of co-segregating markers in mapping populations and open populations, isolation of the complete gene. In a more advanced stage, the procedure is the validation of candidate genes, which will be discussed below.

THE ISOLATION OF GENES FROM THE FUNCTIONAL GENOME

The techniques of functional genomics are expanding and the volume of genetic information is constantly being expanded. Data from the expressed transcriptome (mRNA) constitute a tool that facilitates rapid access and cloning of genes of interest with less effort, when compared to the techniques of genetic mapping. A disadvantage of reverse genetics techniques is the lack of information of regulatory regions that are not detected using the mRNA, the promoter, and the introns. Polymorphism is much higher in these two regions, since it can affect transcription levels by reducing the expression. In the coding region, the occurrence of bases alteration can alter the amino acid sequence of the encoded protein and derail protein activity by gene silencing. Events of pre transcriptional regulation such as epigenetic effects also interfere with the detection, complicating subsequent analysis, even those based on structural genome. Eukaryotic genes are highly influenced by the environment causing large variation in transcription level, and they may lead to erroneous conclusions among contrasting phenotypes. All the techniques of this class depend on other genomic strategies (DNA-based) to obtain complete gene cloning (promoter+regulatory and coding region). A major advantage of these techniques is the fact that, initially, it does not depend on large populations, only on contrasting genotypes that clearly show the stable phenotype (low variation of phenotypic expression for a particular group of genotypes).

CHOOSING A TECHNIQUE FOR FUNCTIONAL GENE CLONING

Techniques based on differential expression may allow us to reach the gene of interest more easily and quickly, but it's necessary to have a broad knowledge of the phenotypic expression character. Often, phenotypic expression is the result of a signaling chain trigger between several genes. In this case, many genes are differentially expressed, and it is necessary to identify the key-gene(s) that control the regulation of this activation; without it the modulation of gene expression is indifferent. Therefore, they are genes that most effectively contribute to a given phenotype. Before starting the activities, it is necessary to properly plan the tests; to promote a controlled induction with a minimal of unwanted interference and to select the correct technique(s) to be used. The techniques require, in most cases, *in silico* analysis (bioinformatics) and the attainment of cDNA or of hybridization processes. Techniques widely used for the detection of differential expression of candidate genes for the control of the phenotypic expression are described in Table 6.4.

Many other techniques such as Real Time PCR are used, which is a powerful tool for both genotyping and gene expression analysis to medium scale. It should be emphasized that these techniques are most effective when combined with others, for example the full-length cDNA subtracting by PCR and the application to RNASeq. Only differentially expressed genes are detected,

TABLE 6.4 Technical Application for Isolation of the Plant Gene Based in Functional Genome

Techniques	Application in the Gene Isolation	Main Advantages or Limitations	Comments
ESTs library	Partial analysis of the transcriptome, important for the cloning of genes from cDNA	Does not discriminate differentially expressed genes and genes expressed much affect data analysis	Genes are mostly fragmented, but the genetic material is stored *in vivo* and may be appealed anytime
Full-length library	More recommended for the isolation of complete sequence of the mRNA	Technique that requires good quality of the mRNA and experienced professional	Ideal for the isolation of intact genes and full-length transcripts
Subtractive and Suppressive Library	Analysis of differentially expressed transcript	Requires fragmentation and amplification of the genetic material which increases the error rate	Enriches the pool of differentially expressed genes limited to analysis in a reduced genome
Northern-Blot	Sequence specific isolation	Requires knowledge of the genes of interest and specific probes	Widely used for the isolation of genes with high fidelity
RACE	cDNA terminal extension	Laborious and unprofitable	High cost for gene cloned
SiteFinder-PCR	Length of major fragments at once both the 5′and 3′ of the DNA	PCR requires, there may be errors, so it needs more coverage in sequencing	Ease of cloning large tracts each time, in most cases, can generate a complete gene only in one step
Functional Markers	Genetic mapping	Can suffer effects of the environment is made from cDNA	The use of reduced genomes promotes co-segregation of markers
Micro arrays	Analysis of gene expression and molecular markers	High cost and need for fine-tuning techniques	Generating information on a large scale
RNASeq	Analysis of transcripts in large scale	Need good knowledge of bioinformatics	It is a technical scale, can generate gene expression analysis of the transcriptome level, and detects polymorphism SNP
Comparative Genome (Gene Synteny)	Cloning of genes not identified or characterized for the species	Requires extensive knowledge of orthologous genes or genetic map that has a highly saturated	The loci are usually kept in order and sequence in some genetically divergent species

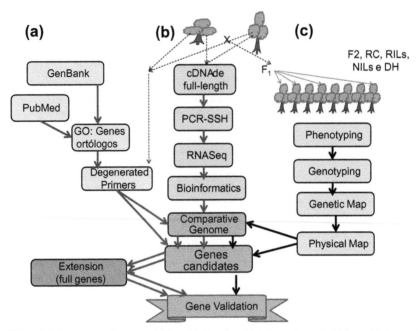

FIGURE 6.7 Systems for gene cloning: (a) Based on bioinformatics and database (b) Suggest a scheme based on reduced functional genome, and (c) Based on structural genome and genetic mapping.

the genome is greatly reduced and significantly expressed genes are subtracted (ribosomal genes), identifying the sequence of each gene. In most cases and depending on the objectives, in which we aim to obtain the complete sequence from the mRNA, or still, it is desired to obtain regulatory regions such as introns and promoter, it is recommended to use the technique of PCR SiteFinder. With it, it is possible at once to clone the promoter region, introns, and transcription termination sequences, however the use of DNA is necessary. In a final stage, the use of PCR with primers specifically designed for regions of interest is required for the isolation of genes, either using techniques of structural or functional genome (Figure 6.7).

THE CLONING OF GENES USING BIOINFORMATICS RESOURCES

The database of genome projects, the NCBI (National Center for Biotechnology Information - http://www.ncbi.nlm.nih.gov/), EMBL-EBI (European Bioinformatics Institute - http://www.ebi.ac.uk/), and DDBJ (DNA Data Banj of Japan - http://www.ddbj.nig.ac.jp/) are the main sources of genomic information resources. NCBI is the database of major importance for the isolation of genes (see Chapter 9: Tools for Bioinformatics).

It is recommended to do constant research on these databases before the preparation of projects in order to detect if something is being done in a correlate manner; during execution, in order to share information with groups of common interests, and after completion, with the intention of establishing guidelines to decide what should be published. This update needs to be constant, because considerable amounts of new information are daily deposited in these databases.

Genes deposited in these databases, from the same species or related species, can compose cloning strategy. The use of comparative genome technology may include small genomic regions (microsynteny) or large ones, such as an entire chromosome (macrosynteny), in a study of genome colinearity. It is necessary to decide why, what, and how to compare, and for that, usually, orthologous genes are used. Orthologous genes come from different species, but they have a common ancestor and the encoded protein retains its properties and functions. Homologous genes can also be used (genes that have a common ancestor, reflected in a high conservation rate of its DNA sequence; genes from different types of plant), paralogous ones (genes that have a common ancestor and that exist in a same genome, but with different functions), analogous ones (genes in different species, but with the same sequence and function), or pseudogenes (genes of high similarity in the sequences, but not functional). In many cases, the attainment of all protein sequences of all types, especially orthologous ones, is aligned, and then degenerate primers are obtained for the conserved sequences. The amplification in parents or contrasting genotypes is performed for the trait and amplicons of interest are selected to start the works.

Synteny is the conservation of genes location in equivalent positions in related species, in gene content and in the order of genomes chromosomes. Macrosynteny aims the comparative study on a long segment of the chromosome, or of the genome, highly conserved in order, whereas microsynteny is based on sequence conservation of large extensions of genomic DNA. There is a great similarity between cereal genomes in related species, separated by a long time of evolution. For example, the 12 rice chromosomes are aligned with the 10 maize chromosomes, the 10 sorghum ones, and the 7 basic wheat and barley chromosomes so that the sequences are highly similar and positioning follows the same order for the genes.

A tool of NCBI for research is the MapView (http://www.ncbi.nlm.nih.gov/projects/mapview/), which is detailed gene mapping, constitutes an important strategy for gene cloning and even of regions of interest. In plants, currently, there are 74 species with their loci-saturated chromosomes that are properly positioned. Two strategies can be applied: (1) the search for markers associated with the region when there is partial knowledge of the locus of interest or (2) the pursuit of a region when there is one or several markers related to the locus of interest in orthologous regions. This reduces the time and cost of localization of the loci of interest. Another important tool, whether the isolation technique(s) of gene(s) of interest is based on functional or structural genome, is the identification of candidate genes. A number of programs are used for this purpose and they should be selected according to the degree of the researcher's knowledge and, essentially, to the rigor of analysis. Gene finding is one of the most stringent stages (Schweikert et al 2009)—the determination of regions encoding genes with their probable ORF (Open Read Frame: transcription initiation region).

VALIDATION OF CANDIDATE GENES

The final stage of gene cloning is validation—the confirmation that gene(s) encode such phenotypic trait. This stage is essential both for the application in breeding and in genetic transformation of plants. Among several techniques, the most applicable ones are:

Phenotypic reversion: It is a practice that consists of cloning the gene with its promoter, or a compatible constitutive promoter, and the transformation of recessive genotypes for the gene in question. In practice, the use of mutant database is an alternative for the selection of these genotypes. TAIR (The Arabidopsis Information Resource: http://www.arabidopsis.org/) with its information base on mutants, the ABRC (Arabidopsis Biological Resource Center), reproduces and distributes seeds and DNA of 4011 mutant accesses of *Arabidopsis thaliana* and related species, for 3,126 defective loci. Other mutant databases are available, such as rice, maize, and soybean ones, among others. Mutants such as *Saccharomyces cerevisiae* yeasts are also particularly used for analysis of transporter genes or genes related to certain amino acids or nutrition of plants. Advantageously, it shows speed of propagation, heterologous expression of enzymes, and its easy attainment for other auxiliary studies.

Interference RNA: In some cases, transformation of mutants does not meet research objectives. The use of genetic transformation by RNAi (interference RNA) in its various techniques is recommend. Among the most indicated is the use of miRNAa (artificial microRNA), in which the construction is based on PCR and the RNAi is specific only to the gene under study. Since it explores the variability among the other alleles of the gene, it is necessary to know allele polymorphism in question, on which miRNAa is drawn. Plants carrying the phenotype are transformed and if gene silencing is detected by the failure in phenotype expression, the gene is validated by qRT-PCR.

Test on mapping population or segregating population: In possession of the detected polymorphism, both in the gene or in regulatory regions, specific primers are established and PCR adjustments are performed in extremely contrasting phenotypes. Then, the analysis in mapping populations is applied, being expected that only genotypes carrying the sequences under study present amplification. As a disadvantage, the technique is not highly efficient in detecting the gene, since in specific cases more than one gene can co-segregate with the phenotype.

Transient Expression: In cases in which the aforementioned methods are not possible to be applied, agroinfiltration for the transient expression of specific genes is an alternative. The technique is based on microinjection of artificial chromosomes, artificial mRNA, microRNAs, or even vectors under the epidermal tissues followed by the induction of the gene in question. In a few hours or days, according to the genes expression, the material is collected for RNA extraction and the analysis of qRT-PCR.

In vitro **heterologous expression**: In more complex cases, it is recommended to normally use gene expression in certain bacteria competent yeast or strains. The induction of the gene of interest is promoted and enzymes, proteins, or metabolic for biochemical analysis are isolated. Still in cases in which the gene is involved in processes of protein/protein or protein/DNA interaction, electrophoretic mobility assays are well accepted.

PROSPECTS FOR THE ISOLATION OF GENES OF INTEREST

The development of genetic maps with markers of easy attainment and of large-scale, highly reproducible, co-dominant and specific to linkage groups is highly desirable for their application in plant breeding. The development of high density maps that incorporate derivatives EST markers, especially SNPs, or yet of new statistical and genetic approaches such as the Genome-wide association study (GWAS, see Chapter 4) would provide greater accuracy in identifying genes or QTLs associated with economically important traits. With the increasing use and ease of access to next-generation sequencing technologies and genotyping in large-scale, strategies that use reduced genomes, genes differentially expressed and RNASeq will assume a key role in the rapid development of genetic maps and physical maps for most cultures of interest. The combination of techniques of functional and structural genome in large scale is essential to the isolation and cloning of genes or QTL even in small laboratories. Soon, the genome sequencing will be done in a few hours and physical and genetic maps will be widely enriched. However, it is important to highlight that the information generated and available in databases has been underutilized so far. It is possible that these bioinformatics tools become part of plant breeding planning as an essential component increasing the production capacity to obtain new cultivars.

Genotyping, in the analysis of DNA polymorphism and of genetic variability of genes is a task that is simplified, but phenotyping still remains one of the bottlenecks, especially regarding large populations. The large scale analysis, which requires computational methods for their implementation and the challenges imposed by the need to compare large, diverse, and complex volumes of data coming from the numerous genome projects have stimulated the development of new methods, algorithms, and bioinformatics tools and also the improvement of the existing techniques. In this sense, it is necessary that the formation of human resources deals with the training of professionals who are able to integrate the various areas of classical breeding science, such as breeding systems and biotechnology as the comparative analysis of complete genomes and transcriptomes.

REFERENCES

Diola, V., Brito, G.G., Caixeta, E.T., Maciel-Zambolim, E., Sakiyama, N.S., Loureiro, M.E., 2011. High-density genetic mapping for coffee leaf rust resistance. Tree Genetics & Genomes 1–10 http://dx.doi.org/10.1007/s11295-011-0406-2.

Ferreira, A., da Silva, M.F., da Costa e Silva, L., Cruz, C.D., 2006. Estimating the effects of population size and type on the accuracy of genetic maps. Genetics and Molecular Biology 29, 182–192.

Haldane, J.B.S., 1931. The cytological basis of genetics interference. Cytologia 3, 54–65.

He, P., Li, J.Z., Zheng, X.W., Shen, L.S., Lu, C.F., Chen, Y., Zhu, L.H., 2001. Comparison of molecular linkage maps and agronomic trait loci between DH and RIL populations derived from the same rice cross. Crop Science 41, 1240–1246.

Kurata, N., Nagamura, Y., Yamamoto, K., Harushima, Y., Sue, N., Wu, J., 1994. A 300 kilobase interval genetic map of rice including 883 expressed sequences. Nature Genetics 8, 365–372.

Kosambi, D.D., 1944. The estimation of map distances from recombination values. Annals of Eugenics 12, 172–175.

Michelmore, R.W., Paran, I., Kesseli, R.V., 1991. Identification of markers linked to disease-resistance genes by bulked segregant analysis: a rapid method to detect markers in specific genomic regions by using segregating populations. Proceedings of the National Academy of Sciences 88, 9828–9832.

Schweikert, G., Behr, J., Zien, A., et al., 2009. mGene.web: a web service for accurate computational gene finding. Nucleic Acids Research 37 (Web Server issue): W312–6. doi:10.1093/nar/gkp479.

Schmidt, R., West, J., Love, K., Lenehan, Z., Lister, C., Thompson, H., Bouchez, D., Dean, C., 1995. Physical map and organisation of *Arabidopsis thaliana* chromosome 4. Science 270, 480–483.

Semagn, K., Bjornstad, A., Ndjiondjop, MN., 2006. Principles requirements and prospects of genetic mapping in plants. African Journal of Biotechnology 5 (25), 2569–2587.

Stam, P., 1993. Construction of integrated genetic linkage maps by means of a new computer package: JoinMap. Plant Journal 3, 739–744.

Tanksley, S.D., Ganal, M.W., Martin, G.B., 1995. Chromosome landing: a paradigm for map-based gene cloning in plants with large genomes. Trends in Genetics 11, 63–68.

Chapter 7

Tissue Culture Applications for the Genetic Improvement of Plants

Moacir Pasqual, Joyce Dória Rodrigues Soares, and Filipe Almendagna Rodrigues
Federal University of Lavras, Brazil

INTRODUCTION

The culturing of cells, protoplasts, and tissues is one of the most successful areas in the realm of biotechnology. The manipulation of cells and cellular components, tissue handling, large-scale and high-speed seedling production, and the adoption of techniques for the genetic improvement of plants are just some of the many important areas of tissue culture. Advances in understanding the physiology of growth and *in vitro* development have led to the optimization of methods for accelerating and improving plant development.

The application of tissue culture methods to the genetic improvement of plants facilitates the development of specimens exhibiting the most advantageous adaptations to various biotic and abiotic stresses, enabling selection pressure in the culture environment itself and saving time and space for the breeder. Furthermore, tissue culture methods can produce numerous benefits in conventional breeding programs by offering a method to overcome the limitations of the reproductive system.

An enormous advantage of tissue culture is the possibility the technique offers to plant breeders for the exploration of genetic diversity within a short period, enabling the concurrent regeneration and proliferation of plants through micropropagation using methods including somaclonal variation, mutagenesis, embryo rescue, the production of double haploid lines, synthetic seed production, *in vitro* selection, and the conservation and exchange of germplasm.

SOMACLONAL VARIATION

The growth of plant cells *in vitro* and their regeneration into plants is an asexual process involving only mitotic divisions, and the expected result is the proliferation of genetically uniform plants. This expectation is essential in the micropropagation industry and provides a technical basis for genetic manipulation in plants. The occurrence of variation during the culture process is not expected and even less desired. Therefore, the tissue culture of plant cells is not expected to provide a useful source of variation for plant breeding.

Evidence of chromosomal instability in cells gained significance when regeneration systems were established from the explants and protoplasts of important crops in which genetic and phenotypic variations were demonstrated. Interest in this phenomenon increased with the widespread reports of changes in agronomic traits and promises of a new source of variability for plant breeding. In 1981, Larkin and Scowcroft termed this variation phenomenon "somaclonal variation," a term now used worldwide.

Origins and Causes of Somaclonal Variation

Regeneration from somatic cells is possible thanks to the phenomenon of totipotentiality. During regeneration, already differentiated cells become undifferentiated and redifferentiate later through a new developmental pathway. Changes occurring during this process take longer in cultured protoplasts. The causes and origins of somaclonal variation are often related to internal and external issues in the plant tissue cultures.

The meristem is the only plant tissue culture system in which the organization remains unchanged during the culture process; therefore, this tissue does not exhibit somaclonal variation. Other systems in which regeneration is achieved through the formation of adventitious meristems after the growth phase of disorganized callus or cell suspensions are vulnerable to instability. The longer the duration of the disorganized growth phase, the higher the likelihood of somaclonal variation. This holds true when regeneration begins in embryogenic or organogenic structures, although the direct formation of these structures from grown tissue (without any intermediate callus phase) clearly minimizes the likelihood of instability. A study performed by López et al. (2010) in cacao tree (*Theobroma cacao*) investigated the possible genetic and epigenetic variations in plants derived from callus cultures. According to the study, the callus cultures demonstrated increased divergence (both genetic and epigenetic) compared with plants derived from organogenesis.

Variation in Tissue Origin

Genomic variations, including endopolyploidy, polyteny, and the amplification and reduction of DNA sequences, can occur during somatic differentiation in plant growth and development. Therefore, unsurprisingly, differences in the frequency and nature of somaclonal variation can occur when regeneration results from tissues of different origins. Furthermore, the processes of differentiation and redifferentiation may involve both qualitative and quantitative genome variations, and different DNA sequences can be expanded or deleted during these cellular variations. The variations are related to the origin of the tissues and the regeneration system, which may explain why certain culture systems are associated with specific types of variations, including the high frequency of albino plants and the expansion of repetitive sequences in anther cultures of cereals and tobacco, respectively.

Experiments in somaclonal variation in which the identity of each somaclone and its donor plant are known, and certain plants regenerated from the same culture can be compared, have demonstrated clearly that although most variations originate from the culture phase, a number of the variations emerge in the donor tissue, indicating that somaclonal variation may originate from somatic mutations already present in the donor plant. Sato et al. (2011) used real-time PCR (polymerase chain reaction) to perform a quantitative study of mutant cells from different parts of plants of the Saintpaulia genus. The study demonstrated that the petals exhibited the highest mutation rates, followed by the sepals and leaves, whereas the stamens and pistils exhibited relatively low mutation rates.

Abnormalities in Cell Division *in vitro*

Mitosis is induced when cells or differentiated tissues are isolated and cultured *in vitro* in the presence of exogenous growth regulators. Cytological studies have demonstrated that abnormalities can occur during cell division, resulting in numerical and structural chromosomal abnormalities in regenerated plants. Growth regulators, temperature, light, osmolarity, and agitation of the culture medium affect the cell cycle of plants *in vivo*, suggesting that inadequate *in vitro* cell cycle control is one of the causes of somaclonal variation.

The problems with cell cycle control are aggravated during protoplast regeneration. When the cell wall is removed, new walls are formed and division is induced in cells derived from protoplasts, and a high frequency of errors has been demonstrated during microtubule synthesis, spindle formation and orientation, and chromatid segregation, resulting in broad variations in the structure and number of chromosomes. Shiba and Mii (2005) investigated the regeneration of plants derived from cultures of *Dianthus acicularis* ($2n = 90$) protoplasts and observed that most seedlings exhibited normal phenotypes, although a number of seedlings demonstrated morphological variations with a reduced number of chromosomes, early flowering, and vigorous growth compared to normal seedlings with a tetraploid chromosome number.

Importance of Growth Regulators

Growth regulators affect the cell cycle and consequently contribute to the causes of somaclonal variation. A high concentration of 2,4-D (2,4-dichlorophenoxy acetic acid) has been implicated in the increase in chromosomal instability. Evidence of the direct mutagenic effect of growth regulators is occasionally contradictory and indicates an indirect effect through the stimulation of rapid and disorganized growth. There is also evidence that the type and concentration of growth regulator affect the variation observed in plants regenerated through the selective stimulation of cells exhibiting specific ploidy levels.

Growth regulators can also cause transient changes in phenotype, and although these changes are inherited mitotically during plant growth, they are not sexually transmitted and are therefore epigenetic. Gas exchange may be inadequate in tissue cultures performed in sealed-cap flasks, and the accumulation of growth regulators, including ethylene, can result in epigenetic modifications. Although not sexually heritable, these variations are relevant in the micropropagation of ornamental species in which the direct production of the tissue culture is important for species that are propagated vegetatively or in plants with long reproductive cycles.

Bairu et al. (2006) investigated the effect of the type and concentration of plant growth regulators and subculture on somaclonal variation in Cavendish banana plants. The study used plants grown in the fourth cycle of multiplication in the presence of auxins (indole-3-acetic acid, IAA; indole-3-butyric acid, IBA; and 1-naphthaleneacetic acid, NAA) and cytokinin phytoregulators (6-benzylaminopurine, BAP; and thidiazuron, TDZ) used for multiplying shoots for 10 generations. The relationship between the rate of multiplication and somaclonal variation was evaluated by correlation analysis. The data demonstrated that the treatments with higher rates of multiplication produced more variants (up to 72%).

Responses of Plant Genomes

The forced changes in the differentiated state of plant cells during the culture and regeneration process and the particular conditions of the culture environment generate conditions of stress for the plant genome. Therefore, somaclonal variation can be regarded as a stress-induced tissue culture variation. Different genomes respond differently; therefore, somaclonal variation has a genotypic component in that certain genotypes are stable in tissue culture, whereas others are consistently unstable.

Hossain et al. (2012) evaluated the morphological and genetic variations in chili pepper (*Capsicum annuum* L.) somaclones derived from tissue culture (Figure 7.1). Cotyledon explants of Shishitou and Takanotsume cultivars were cultured in Murashige and Skoog (MS) media supplemented with $5 \, mg \, L^{-1}$ benzyladenine (BA) for regeneration, and the shoots were rooted in MS medium supplemented with $0.1 \, mg \, L^{-1}$ NAA and $0.05 \, mg \, L^{-1}$ IBA. The regenerated plants (R0) were selfed (self-pollinated) to produce seeds for the next generation (R1 lines). The qualitative traits were investigated in the R0 generation, and the qualitative and quantitative traits were investigated in the R1 generation. In the R0 generation, changes were observed in the growth characteristics of the plants and in the color of the stem, flowers, and fruits (anthocyanin expression). The morphological and agronomic traits were compared in the R1 and the parental lines. A significant variation within the R1 lines and differences between the

FIGURE 7.1   Somaclonal variation in trait qualities. Green stem (A), white flower (C), and green fruits (E) of the parental cultivar Shishitou. Purple stem (B), purple flower (D), and purple fruits (F) of the SR0-P regenerated plant, and R1 line dwarfism, SR1-2 (G). (*Source: Hossain et al., 2012*).

R1 and parental lines were noted. Genetic variations between three somaclones were demonstrated using random amplification of polymorphic DNA (RAPD) analysis. Variations including early flowering and increased yield components are indicative of the response to selection for any plant. The occurrence of productive variants between somaclones of cultivars, including the Shishitou and Takanotsume cultivars, indicates the possibility for their improvement through somaclonal variation.

The correlation between chromosomal breaks in tissue cultures and the presence of heterochromatin has been demonstrated in several studies. This correlation can be explained by the late replication of heterochromatin, resulting in "cell cycle disruption," which affects the timing of chromosome replication during the S-phase; therefore, cell division would most likely result in chromosomal aberration. Based on this theory, plants with chromosomes exhibiting increased amounts of heterochromatin are vulnerable to higher frequencies of chromosomal aberrations, which could explain the differences in stability between different species. However, other genome components may be involved because differences in somaclonal variation are observed between closely related species and between different cultivars of the same species.

MUTAGENESIS

Mutagenesis methods have generated interest with the advent of tissue culture techniques. The possibility of handling a large number of totipotent cells, haploid or diploid, combined with the potential for plant regeneration, has generated higher expectations towards attaining desired mutants.

The use of *in vitro* cell cultures treated with mutagenic agents reduces the incidence of chimeras, facilitates the selection of mutants for dominant or recessive traits that appear with a low frequency, and enables the use of chemical mutagens because this *in vitro* method circumvents the difficulty associated with tissue penetration when the treatments are performed *in vivo* in plant tissues propagated vegetatively.

The induction of mutations, both *in vivo* and *in vitro*, is regarded cautiously as a method for obtaining improved plants because the modifications cannot be directed, that is, they occur by chance. The chances of modification increase when the study is performed *in vitro* because a large number of individuals can be regenerated in a limited space. This method is recommended when the expansion of natural genetic variability is desired or when there are restrictions on the use of conventional genetic improvement methods. Garlic has been cited as an example for *in vitro* mutagenesis because the lack of flowering in this plant precludes the use of crosses between different cultivars.

According to Ahloowalia et al. (2004), the main strategy for the use of mutagenesis in genetic breeding is to gain one or two of the traits of greatest interest without excessively changing important agronomic traits, including yield.

Physical and chemical mutagens exhibit advantages and disadvantages and can be used for *in vitro* mutagenesis as follows:

1. Chemical mutagens. Examples of chemical mutagens include:
   - sodium azide (NaN_3), which inhibits the activity of cellular respiration enzymes;
   - antibiotics (streptomycin, mitomycin C, and actinomycin D), which cause chromosomal breaks;
   - base analogs [BU (5-bromouracil) and Bud (5-bromodeoxyuridine)] of similar composition to DNA nitrogenous bases, which cause base-pairing errors in DNA replication;
   - alkylating agents [diethyl sulfate (DES), ethyl methanesulfonate (EMS), methyl methanesulfonate (MMS), among others], which promote alkylation of phosphate groups in reactions with DNA;
   - purine and pyrimidine bases, which are the most reactive mutagens;
   - other compounds (nitrous acid, acridine, hydroxylamine, among others).

Silva et al. (2013) evaluated the mutagenic and antimutagenic and genotoxic activities of extracts from the *Mimosa tenuiflora* species on peripheral blood cells using the Ames and micronucleus tests and concluded that the extract is not mutagenic in the absence of an exogenous metabolic system and does not induce an increased frequency of micronuclei. Under these conditions, the agent was characterized as non mutagenic.

2. Physical mutagens. Electromagnetic radiation (ultraviolet, X-rays, and gamma rays) and corpuscular radiation (alpha- and beta-particles, protons, neutrons, among others) are examples of physical mutagens. The radiation energy is transferred directly or indirectly to the DNA in cells through physical and chemical processes (collision, excitation, and ionization), causing damage to the cellular components that can generate erroneous base replacements during reconstitution, changing the genetic code and causing mutations. Gamma rays are obtained primarily through the radioisotopes 60Co and 137Cs and are the most commonly used physical mutagen because they exhibit increased penetrating power and can reach all the cells in any tissue type. Ultraviolet rays exhibit less penetration power and are used only for the treatment of single cells or small cell clusters.

Ferreira et al. (2009) evaluated the effect of various doses of gamma radiation on explant buds from fig trees of different sizes. The seedlings were established *in vitro*, were separated according to size (2.5 to 4.5 cm, 5 to 7 cm, and 8 to 10 cm), and were irradiated with 10, 20, 30, 40, and 50 Gy. After the irradiation, the seedlings were transplanted into explants containing a single bud, were separated according to bud position (basal, median, or apical), and were grown in woody plant medium (WPM). The following traits were evaluated after 90 days' incubation in the growth room: explant mortality, root formation, shoot length, bud number, and seedling weight. Doses of up to 50 Gy did not cause seedling death, and doses greater than 30 Gy prevented the formation of roots; therefore, the 30-Gy dose was recommended for the irradiation of fig tree seedlings measuring greater than 2.5 cm.

After applying mutagens to the explants, which can be in the form of a callus or cell suspension, mutant cells, such as buds and pollen grains, among others, must be selected, and these cells will multiply and generate plants. Methods have been established for generating certain mutations, including resistance to disease, herbicides, and soil salinity, which incorporate the problem-causing agents into the culture medium, thereby facilitating the identification of resistant or tolerant cells.

PROTOPLAST FUSION

Because protoplasts lack a cell wall, the cell membrane is the only barrier between the extracellular and intracellular environments. Plant somatic hybridization involves four distinct stages: (1) protoplast isolation; (2) protoplast fusion; (3) regeneration of plants from selected tissues; and (4) analysis of regenerated plants. Somatic hybridization provides a means to overcome sexual barriers in plant breeding because technically, any protoplast combination can be induced to undergo fusion. This method not only provides a means to generate hybrids between sexually compatible plants but also facilitates the genetic modification of sterile or subfertile plants of vegetatively propagated species and plants with long reproductive cycles.

Protoplast fusion produces heterokaryons and homokaryons, and a number of the protoplasts remain unfused. The heterokaryons containing nuclei from both of the genera, species, or varieties used initially in the cytoplasm mixing are the fusion product of the intended genetic manipulation and can develop within hybrid cells (Figure 7.2). Hybrid somatic cells are totipotent and can form plants through organogenesis or embryogenesis.

The incompatibility between the two nuclear genomes or the nucleus-cytoplasmic combinations may become apparent, leading to the elimination of chromosomes or, in extreme cases, failure of the heterokaryons to sustain growth and division. Certain treatments, including the irradiation of protoplasts, can increase the elimination of chromosomes, resulting in fusion products that retain the nuclear genome of one parent in a mixed cytoplasm (cybrids). These treatments facilitate the intergeneric and interspecific transfer of extranuclear genetic elements, including mitochondria and chloroplasts, and the treatments are used to facilitate the transfer of genes that control chlorophyll content, resistance to herbicides, and cytoplasmic male sterility. Normally, cytoplasmic genomes are inherited maternally following sexual hybridization. Consequently, new nuclear-cytoplasmic combinations can be produced sexually only through backcrossing, which is a time-consuming and random process, and the cybridization technique reduces this time considerably.

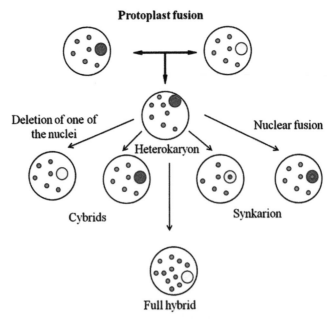

FIGURE 7.2   Diagram of full hybrid formation from protoplast fusion.

Protoplast Fusion Method

Protoplast fusion can be induced using chemical or electrical methods. In both cases, the plasma membranes are disrupted temporarily, resulting in the formation of pores and cytoplasmic connections between adjacent protoplasts. These connections are known to inhibit pore closing and enable lipid molecules randomly oriented in the pores to unite and form membranes between adjacent protoplasts.

Chemical Fusion

High concentrations of chemicals, including polyethylene glycol (PEG), dextran, and polyvinyl alcohol (PVA), have been used to induce pore formation. PEG is the chemical most commonly used for this purpose. PEG with a low carbonyl content, such as 30% PEG 1500 solution, must be used to obtain a high frequency of heterokaryon formation (above 10% of the treated protoplasts) and to ensure viability.

The development of biotechnological tools, including protoplast fusion, provides a means of circumventing the natural barriers of the reproductive biology of various species found in conventional breeding systems. Costa et al. (2003) reported the production of four somatic interspecific hybrids, Cleopatra mandarin + Volkamer lemon, Ruby Blood orange + Volkamer lemon, Rohde Red Valencia orange + Volkamer lemon, and Rangpur lime + Sunki mandarin and an intergeneric somatic hybrid Valencia orange + *Fortunella obovata* (Tanaka), which have been included and evaluated in breeding program rootstocks. To generate the hybrids, protoplasts were isolated from embryogenic calli and leaves and were chemically fused using polyethylene glycol. The plants were regenerated via somatic embryogenesis, and somatic hybridization was confirmed through leaf morphology, cytology, and DNA (RAPD) analyses.

Electrofusion

The following steps are used in protoplast electrofusion: (1) alternating current is used to transfer the protoplasts and to promote a close contact between the membranes, and (2) continuous short pulses are used to induce membrane disruption at the contact points. Despite the need for expensive and sophisticated equipment to generate alternating current fields and continuous current pulses, the electrofusion method has become increasingly popular because it is less damaging to protoplasts than chemical procedures.

Geerts et al. (2008) performed a study using protoplast fusion technology by somatic hybridization in *Phaseolus*. The success of interspecific breeding between *Phaseolus vulgaris* L. (PV) and the two donor species, *Phaseolus coccineus* L. (PC) or *Phaseolus polyanthus* Greenm. (PP), required the use of donor species as the female parents. Although the incompatibility barriers were postzygotic, the success of the phase-F1 cross was very limited given the hybrid embryo abortion. The study described the use of protoplast fusion methods within the genera *Phaseolus* as an alternative to conventional crosses between PV, PC, or PP. A large number of heterokaryons were generated through different genotypes using protoplast fusion procedures based on both electrical (750 or $1500\,V\,cm^{-1}$) and chemical fusion using polyethylene glycol (PEG 6000) as the fusing agent.

Selection of Heterokaryons and Hybrid Somatic Tissues

The most difficult aspect of somatic hybridization is the selection of heterokaryons, cells, or tissues derived from heterokaryons or hybrid plants. In a number of studies, the protoplasts subjected to fusion treatments were cultured, and the regenerated plants were selected based on their traits. This procedure is laborious and takes up a large amount of space.

When two populations of protoplasts are easily identified, individual heterokaryons can be separated using micropipettes. This method works well for leaf protoplast fusions, which contain chlorophyll, and discolored protoplasts can be isolated from cell suspensions. Great advances have been made in this field with the development of the current system of fluorescent labeling in heterokaryons. The protoplasts are stained green through treatment with fluorescein diacetate (1–$20\,mg\,L^{-1}$) and are fused with protoplasts emitting red fluorescence, derived from autofluorescing chlorophyll and exogenously applied rhodamine isocyanate, respectively.

A number of procedures inhibit the growth of homokaryons of at least one of the parental populations. Hormone autotrophic, auxotrophic, and antibiotic resistant mutants, amino acid analogs, and fungal toxins have been used in various combinations. Albino mutants have also been used successfully, and hybrid tissues have been identified through the production of chlorophyll following exposure of the cultures to light.

Characterization of Somatic Hybrid Plants

A number of somatic hybrids have been morphologically, cytologically, and biochemically characterized. Routine biochemical characterization includes the analysis of isoenzymes and proteins, plant resistance to viral infection and antibiotics, and sensitivity to herbicides or fungal toxins. More recently, restriction fragment length polymorphism (RFLP) and RAPD analyses have been used.

Patel et al. (2011) developed somatic hybrid plants of *Nicotiana* × *sanderae* (+) *N. debneyi* that were resistant to *Peronospora tabacina* using the protoplast fusion method. Somatic hybridization was used in the study to generate a new hybrid combination involving two sexually incompatible tetraploid species. The somatic hybrid plants were characterized using molecular, cytogenetic, and phenotypic approaches.

EMBRYO RESCUE

An application of tissue culture that has helped many plant breeders is rare embryo rescue. Many genes that confer tolerance or resistance to biotic or abiotic stresses have been lost because of the intense process of domestication imposed by humans on species of cultivated plants. Therefore, a consequence of the reduced genetic variability is the difficulty breeders experience in locating sources of resistance or tolerance in accessions of the same species, making it necessary to resort to wild species to isolate new tolerance and resistance genes. Although sexual crossing between different species is possible, in gamete fusion and embryo formation, embryo abortion (because of endosperm malformation) as a post-zygotic barrier is common.

Embryo rescue in culture media promotes normal development, enabling the generation of interspecific hybrids and in some cases hybrids between different families, which greatly facilitate the breeder's search for genes of interest.

The culture medium must contain all the nutritional requirements to replace the endosperm function of nourishing the embryo. Depending on the developmental stage, the embryo should require only inorganic nutrients and a carbohydrate source in the culture medium, although supplementation with growth regulators, antioxidants, vitamins, and other substances is required for very young embryos.

Promising results have been observed using the zygotic embryo culture method for the propagation of palm trees, demonstrated by the increased germination rate, plant uniformity, and conversion of viable seedlings in species including *Cocos nucifera* (Pech-Aké et al., 2007), *Astrocaryum ulei* (Pereira et al., 2006), and *Hyophorbe lagenicaulis* (Sarasan et al., 2002). In contrast to what was observed for a number of commercial species of palm trees, such as *Cocos nucifera* (Ledo et al., 2007), very little is known regarding the correct salt concentration and supplementation required in the *in vitro* environment during the acclimatization phase to obtain vigorous seedlings with higher survival rates.

Soares et al. (2011) evaluated the effect of different concentrations of mineral salts and coconut water on the *in vitro* germination of macaúba palm (*Acrocomia aculeata*) seedlings, which take approximately 2 years to germinate in nature. Embryos were excised and were inoculated into test tubes containing 15 mL of MS culture medium at 50 and 100% concentrations of mineral salts, supplemented with coconut water (0, 50, 100, and 150 mL L^{-1}). The cultures were maintained in a growth room at approximately $42\,W\,m^{-2}$ irradiance, $25 \pm 2°C$, and a 16-hour photoperiod. A higher percentage of embryo germination was observed at 60 days in the MS medium at the original concentration of salts (95.6%). The growth and conversion of viable or normal seedlings that were acclimatized required MS culture medium containing half the salt concentration supplemented with 50 mL L^{-1} coconut water.

Applications of Embryo Cultures

Nonviable Hybrid Embryo Rescue

The hybridization process involves a sequence of events that includes pollen germination, pollen tube growth, fertilization, embryo and endosperm development, and seed maturation. There are a number of barriers to hybridization, which can be classified as pre fertilization (geographic isolation, apomixes, and pollen-pistil incompatibility) and post-fertilization (different ploidy levels, chromosome alterations, elimination of chromosomes, incompatible cytoplasms, seed dormancy, and embryo collapse) barriers.

Interspecific and intergeneric crossings offer plant breeders a method to increase genetic variability and the transfer of desirable genes between species, mainly from wild to cultivated species. Barriers to pre- and post-fertilization can occur during the crossings, resulting in wilting seeds and abortive embryos. For example, pollen failure during penetration in an odd pistil or in two distantly related genomes may be incapable of producing a viable embryo when combined. However, the use of hybridization between closely related species is often limited by failures in post-fertilization endosperm development, that is, fertilization occurs and the embryos begin to develop but degenerate before reaching maturity given the inability of the endosperm to supply the embryos with nutrition. Therefore, hybrid embryos can be rescued if they are removed before the abortion occurs and are artificially cultivated on a nutrient medium.

According to Alves et al. (2011), the African oil palm tree (*Elaeis guineensis*) is the world's greatest source of vegetable oil. However, a fatal yellowing disease is decimating crops in the state of Pará, Brazil. Interspecific hybrids of *Elaeis oleifera* × *Elaeis guineensis* are a viable alternative to overcome this problem. The study demonstrated that different varieties require different concentrations of 2,4-D for callus induction, and the appropriate concentrations of 2,4-D were 375 and 625 μM for the SJ-167 and SJ-165 varieties, respectively.

Overcoming Seed Sterility

A number of species produce sterile seeds that fail to germinate. This infertility can result from incomplete embryo development, mutations in structures covering the embryos resulting in their death, or a type of recalcitrant dormancy for which no method of breaking dormancy has been developed. Methods of embryo culture can produce viable "seedlings" from the seeds.

Macaúba palm is a highly productive oil palm tree, and the oil from this tree can be used in biofuel production; however, the seeds take approximately 2 years to germinate in nature. Therefore, Soares et al. (2011) investigated the effect of various concentrations of salts and coconut water in the MS culture medium on the *in vitro* germination of zygotic embryos of the macaúba palm. The highest germination percentage was observed at 60 days on MS medium containing the original concentration of salts.

PRODUCTION OF DOUBLE HAPLOID LINES

The production of superior lines for hybrid formation requires a long period to reach homozygosity, which is typically reached after seven to nine generations of selfing, corresponding to between 7 and 9 years of work for the breeder (Borém and Miranda, 2009). This timeline refers to annual species; the expected time to produce homozygous lines is even greater for perennial species.

A method to accelerate the production of homozygous lines is through haploid production followed by chromosome duplication, in a process termed haplodiploidization. Using this method, populations of lines in perfect homozygosity (with all their loci in homozygosis) can be produced in just one generation, starting with the F1 population (Figure 7.3). This timeline corresponds to a significant gain in breeding time, greatly facilitating the production of new cultivars. Between seven and nine generations of selfing are required to reach homozygosity using the conventional method; however, with haplodiploidization, lines with 100% homozygosity are obtained after the F1 generation, providing savings in time and costs.

According to Germanà (2011), the biotechnological method to produce haploids and double haploids through gametic embryogenesis has been recognized as an auxiliary tool in plant breeding. Haploids are sporophytic plants with a gametophytic chromosome number, whereas double haploids are haploid plants that have undergone spontaneous or induced chromosome duplication.

There are several methods for producing haploid plants, and anther culture is the most efficient and fastest of these methods. Haploid plants containing half the number of chromosomes characteristic of the species are of interest to genetics and breeding. The plant material is free from the issues of dominance and recessiveness because it contains only one allele at each locus and enables the rapid production of homozygous plants by duplicating the number of chromosomes in a single step, replacing several generations of selfing that are required in the process usually used for producing lines.

Homozygosity in haploid seedlings is achieved immediately after the duplication of the chromosome number, which can be spontaneous or induced by chemical agents, including colchicine; each chromosome will have an exact copy, and colchicine recovers the diploid condition and restores fertility.

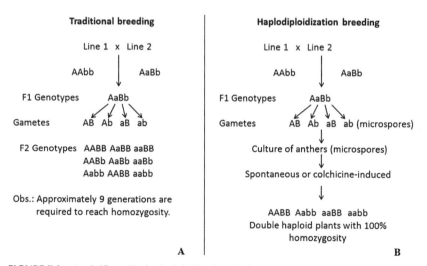

FIGURE 7.3 A – Selfings; B – haplodiploidization of microspores.

Haploid plants can be produced both *in vivo* and *in vitro*. Haploid embryos can be produced spontaneously *in vivo* by delayed pollination, pollination using irradiated or abortive pollen, heat shock treatment of pollen, distant hybridization, and other methods. Haploid plants are produced by *in vitro* culture through the development of unfertilized egg cells, eliminating one of the genomes by culturing young embryos in hybrids resulting from crosses between genetically very distant specimens and most commonly by culturing anthers and/or pollen grains.

Upon producing haploid plants from an anther culture, the next step is to reestablish the diploid condition, which is now in homozygosis. The methods used for the duplication of chromosomes in haploids are varied; however, treatment with colchicine or endomitosis is generally used.

Colchicine inhibits spindle fiber formation, is generally used in aqueous solution or dissolved in alcohol or glycerin, and can be applied as a lanolin or agar paste. The optimum concentration of colchicine varies with the species but usually ranges from 0.1 to 0.5%. However, a number of limitations are associated with the use of colchicine, including mutagenicity, loss of vigor in certain species, and the appearance of plants with various ploidy levels. Endomitosis is the duplication of chromosomes without the duplication of the cell nucleus. This process is common in haploid calli and can be used for producing homozygous diploid plants.

Rodrigues et al. (2011) demonstrated a lower survival rate in explants following exposure to 5.0 mM colchicine for 24 and 48 hours. In the study, two diploid banana plants (1304-04 [Malaccensis × Madang (*Musa acuminata* spp. banksii)] and 8694-15 [0337-02 (Calcutta × Galeo) × SH32-63]) were investigated to establish the effect of different concentrations and different exposure times of colchicine and amiprophos-methyl (APM) on chromosome duplication.

The cultured cells repeatedly demonstrated cytological instability with a tendency towards increases in the ploidy level. The identification of haploid plants is partly possible using marker genes related to color or morphological traits, preferably traits that manifest in the seeds or "seedlings." However, a clear assessment of the ploidy level is only possible through chromosome counts.

SYNTHETIC SEED PRODUCTION

The original concept of synthetic seeds was introduced by Murashige (1978), describing the seed as "an encapsulated somatic embryo"; however, the concept currently accepted is that of Aitken-Christie et al. (1994), which defines synthetic seed technology as "somatic embryos artificially encapsulated, sprout or other tissues, that may be used for *in vitro* or *ex vitro* culture." According to Redenbaugh et al. (1986), the development of synthetic seeds is a plant tissue culture method, which consists of encapsulating somatic embryos, stem apexes, and axillary buds. Therefore, all types of propagules can be included in the definition as structures that can be used in a manner similar to botanic seeds. Figure 7.4 shows an outline of the development of encapsulated explants of

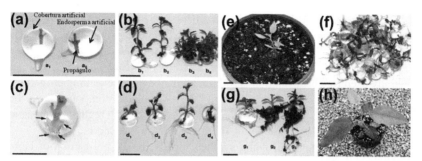

FIGURE 7.4 Development of encapsulated explants of *Corymbia torelliana* × *C. citriodora* under aseptic and non aseptic conditions. (a) Capsules formed by apex (a1) or nodal segment (a2) encapsulation. (b) Regenerated sprouts derived from capsules after 4 weeks in different culture media (b1, 1/2 MS; b2, MS; b3, MS + 2.2 μM BA; and b4, MS + 2.2 μM BA + 0.3 μM NAA). (c) Roots (arrows) induced in synthetic seeds. (d) Seedlings formed directly from sprouts after pretreatment with AIB at different levels (d1, 0; d2, 4.9; d3, 19.6; or d4, 78.4 μM). (e) Potted seedlings after 2 weeks of acclimatization. (f) Synthetic seeds after 4 weeks of pre conversion. (g) Seedlings formed directly from synthetic seeds after pre conversion and planting in different non sterile substrates for 4 weeks (g1, vermiculite and perlite; g2, substrates; g3, organic compound). (h) Potted seedlings at 2 months of age. Scale bar = 1 cm. (*Adapted from Hung and Trueman, 2012*)

Corymbia torelliana × *C. citriodora* from the encapsulation phase to the plant acclimatization phase.

The production of synthetic seeds helps minimize the cost of production and sale of micropropagated plants. Seed propagation has been unsuccessful in most fruit plants because of seed heterozygosity, seed size, the presence of a small endosperm, a low seed germination rate, and the absence of seeds in certain varieties. Many species have seeds that are sensitive to desiccation and can be stored for only a few weeks or months. The interest in using encapsulation technology is increasing for several species given the disadvantages (Rai et al., 2009).

Encapsulation technology offers the potential for the easy handling and exchange of plant germplasm between laboratories, short- or long-term storage, and the direct and efficient transfer of the *in vitro* materials to greenhouses (Rai et al., 2009). The first reports of synthetic seeds described the use of somatic embryos (bipolar structures) (Kitto and Janick, 1982). However, other types of explants can be present in synthetic seeds in the form of unipolar structures, including hairy roots (Uozumi et al., 1992), apical sprouts (Singh et al., 2009), axillary buds (Singh et al., 2010), and protocorms (Sarmah et al., 2010).

The possibility of producing synthetic seeds emerged when somatic embryogenesis began to evolve. The term somatic embryogenesis is used to describe the process of embryo development from somatic cells or haploids. Somatic embryos have a bipolar structure, consisting of the shoot apex and root apex, a closed vascular system with no connection to the initial explant tissue, and their development is characterized as vegetative propagation.

The gel encapsulation system is the most popular of the different methods available. Redenbaugh et al. (1986) pioneered the use of hydrogel, which is similar to sodium alginate, for the production of artificial seeds, demonstrating frequencies of 86% conversion in synthetic seeds of *Medicago sativa*. The gel generally used for encapsulation is sodium alginate, given its gelling properties, low cost, ease of use, and lack of toxicity. Additionally, components can be added to the gel, including minerals, carbohydrates, growth regulators, opaque pigments, among others, mimicking the formation of endosperm.

The main advantages of using synthetic seeds are the production of large numbers of somatic embryos within a short period, requirement of reduced space, maintenance of clonal identity, seed production, independence from seasonal effects, and absence of structures for seedling acclimatization.

IN VITRO SELECTION

Significant improvements in the adaptation of many species have been achieved by the selection and propagation of "superior" specimens. Tissue culture, which greatly accelerates the identification of these specimens, has enabled the investigation of plant-pathogen interactions and the *in vitro* selection of materials resistant or tolerant to the causative agents of biotic and abiotic stresses.

The process of *in vitro* selection usually involves a population of cells that tolerate a situation and develop resistance or tolerance to stress. The goal is the reorganization of the cell into a whole plant exhibiting resistance. This approach assumes that tolerance operates at the cellular level and can act to some degree in the whole plant. The use of tissue culture has the following advantages: the experimental units are maintained in a controlled environment and defined medium, enabling the selection of genotypes that demonstrate resistance to certain stress conditions; cultured cells can be exposed to the selective agent uniformly and individually, reducing the possibility of leaks and undesirable interactions; reduced space is required, which is a distinct advantage over the use of large and costly infrastructures (test fields or automated greenhouses); and absence of external contamination with the causative agent of the disease or other factors because the seeds remain confined to the laboratory.

Research in this area has focused the selection of strains tolerant to salt stress, freezing, herbicides, and resistance to chemical molecules, including metals (aluminum and manganese). Cançado et al. (2009) was the first to address the potential of grapevine rootstocks for resistance to aluminum toxicity. In the study, the primary aluminum target site was confined to actively growing root tips, and the symptoms of early stress in response to chemical elements were measured through the severe inhibition of root growth. The data demonstrated that there was sufficient genetic diversity in the grapevine rootstocks for resistance to aluminum, which enabled the development of breeding programs focusing on increasing the plant resistance to aluminum, particularly in acidic soils of tropical regions where aluminum toxicity is a significant agronomic problem.

GERMPLASM CONSERVATION AND EXCHANGE

In vitro germplasm conservation is an auxiliary tool used in the conservation of genetic resources, especially for vegetatively propagated species. The conservation of germplasm in fruit species is almost exclusively performed in the field, with a small number of duplications of accessions of certain species in the form of orthodox seeds and/or culture of meristems *in vitro* (Ferreira, 2011). This method is a complementary alternative to conventional methods for the maintenance of germplasm banks (Santos et al., 2012). The *in vitro* method of conservation using culture medium for slow growth involves the maintenance of plants in the laboratory under reduced metabolism using periodic subcultures of apical and nodal segments (Souza et al., 2009). This method enables maximization of the interval between subcultures without affecting the viability of the cultures, providing savings in labor costs (Moosikapala and Te-Chato, 2010).

The germplasm is conserved in a physical base, known as the germplasm bank. Germplasm banks are located in research centers or public and private institutions in the form of seeds, *in vitro* explants, or field plants (Ferreira, 2011).

Methods to produce slow growth have been applied successfully to several plant species for genetic resources conservation, enabling the storage of physiologically stable germplasm that exhibits long intervals for subculture *in vitro*, which saves time and reduces maintenance costs. The application of the slow growth method relies on reducing the strength of the culture medium and the temperature and is often combined with a decrease in light intensity or, in some cases, maintenance in the dark (Moosikapala and Te-Chato, 2010).

Currently, there is great interest in maintaining the genetic variability found in nature, which is being reduced gradually with the increasing use of natural resources. Because genetic breeding requires variability for the selection of superior materials for a particular trait, a wide range of genes found in the natural environment must be preserved. *In vitro* culture is useful for maintaining the germoplasm of several species in a small space for a long period. Therefore, plants produced *in vitro* or even plant parts, tissues, organs, and cells can be maintained in culture media with minimum growth or through cryopreservation because both methods enable the material to survive for a long period at a minimal or null growth rate.

The maintenance of germplasm is also important for the conservation of a number of vegetatively propagated plants, including sexually sterile plants (haploids, sterile mutants, male-sterile lines, among others), rare aneuploids, plants with unusual combinations of chromosomes that can be lost through sexual propagation, or even specific heterozygous gene combinations. Preferably, the material to be preserved should be derived from meristems because the meristem is free from viruses and is genetically stable.

An enormous advantage of *in vitro* conservation is germplasm exchange because it enables genotypes to be transported easily and safely between countries, without introducing new diseases in a given country. The plant material imported *in vitro* has the additional advantage of typically reaching its destination under conditions that enable its multiplication, which does not often occur with plant cuttings or other types of propagules given the time that the materials are held at customs.

Slow Growth (Short Periods)

The slow growth method is used for the conservation of meristems of many species because it enables the drastic reduction of plant metabolism without affecting plant viability. This reduction in metabolic activity can be achieved by reducing the light intensity or temperature, increasing growth retardants in the culture medium, or decreasing the concentration of salt and organic components in the culture medium (Razdan, 2003). However, the success of this method depends largely on the physiological characteristics of the species to be conserved, which must exhibit slow growth.

Most protocols for slow growth use the meristems, although various tissues and organs can be used. Genetic alterations are less likely to occur because the meristems are free of pathogens and in most cases are the best explants for micropropagation. Meristematic cells are more resistant to the low temperatures used in the preservation process.

Temperature reduction (10 to 20°C) and nutrient medium supplementation with osmoregulators, such as 4% mannitol, have been used to preserve propagules of clonally propagated species, including tubers, roots, and seasonal fruit trees (Razdan, 2003). The baseline temperature limit allowed for the species to be conserved *in vitro* must be known when using this method.

The slow growth method is not considered safe for cell and callus conservation systems (non organized structures). Although some success has been reported for conservation using these structures, the possibility of somaclonal variation in the cultures must be considered. Furthermore, the cultures are not considered sufficiently stable for long-term conservation. Santos et al. (2012) investigated different culture media and two temperature conditions (18 and 25°C) in vetiver (*Chrysopogon zizanioides*) accessions in a slow-growth protocol for *in vitro* conservation and determined that three vetiver accessions could be preserved in the system for 270 days when the concentration of MS salts was reduced to 25% of its normal concentration and the temperature was maintained at 18°C.

In a study performed using 16 species and subspecies of the genus *Turbinicarpus* (Cactaceae), Balch et al. (2012) demonstrated that supplementing the culture media with osmotic agents, including mannitol ($30\,g\,L^{-1}$) and sorbitol ($30\,g\,L^{-1}$), and lowering the incubation temperature ($4 \pm 0.5°C$) enabled the reduction of *in vitro* growth without affecting the viability rate of the sprouts. The material was stored for 12 months under the treatment conditions, and the seedlings were regenerated in media containing cytokinins after this period. The shoots formed roots, and the generated plants adapted to and survived in the soil, with efficiencies similar to the shoots not treated for slow growth. This method enabled the maintenance of a viable tissue bank for the species *in vitro*, with minimal maintenance, and produced whole plants (Figure 7.5).

FIGURE 7.5 (a) *Turbinicarpus pseudopectinatus* sprouts after 12 months in culture media containing mannitol. (b) *Turbinicarpus schmiedickeanus* subsp. *klinkerianus* sprouts after 12 months in culture media containing sorbitol. (c) *Turbinicarpus laui* sprouts after 12 months in culture media containing mannitol. Production of new sprouts in (d) *T. schmiedickeanus* subsp. *schmiedickeanus*, (e) *Turbinicarpus subterraneusm*, and (f) *Turbinicarpus laui* explants preserved for 12 months in media containing mannitol. Scale bar = 10 mm. (*Source: Balch et al., 2012*)

Cryopreservation (Long Periods)

In the cryopreservation method, the material remains stored in liquid nitrogen at −196°C. The metabolic processes, including respiration and enzymatic activity, are inactivated at that temperature (Benson et al., 1998). The success of a cryopreservation protocol is based on the following six critical factors (Withers, 1985): pre freezing, which starts several days before freezing and involves manipulating the *in vitro* culture conditions to increase tolerance to freezing, supplementing the culture medium with osmoregulating compounds or other additives; cryoprotection, which consists of taking precautions to enable the cells to withstand freezing, the precautions are mainly associated with the toxicity of cryoprotectants; cooling, which consists of decreasing the temperature when preparing the explant for freezing; storage, which is performed in liquid nitrogen at −196°C in the liquid phase or at −150°C in the gas phase; thawing, which is the recovery phase of the stored material; and growth recovery, which evaluates the viability of the material after the period of freezing and preservation.

Cooling can be performed in two ways, slowly or rapidly. In the slow method, the material is frozen gradually, reducing the temperature by 0.1 to 3°C/min. Therefore, a protective cellular dehydration occurs that forms extracellular ice crystals without damaging the cells or causing cell lysis. This slow method is recommended for plants but is dependent on the use of a stepwise-temperature freezer, which is typically very expensive. In the rapid (fast) method, the plant material is placed directly into liquid nitrogen, and the rate of the temperature decrease is in the range of 1000°C/min. In rapid cooling, the main risk is the formation of intracellular ice crystals, which can damage cells and preclude the entire method. Benson et al. (1998) reported that it is usually more difficult to obtain satisfactory results using rapid cooling than slow cooling.

Thawing is as important as cooling (Benson et al., 1998) and may invalidate the cooling, freezing, and storing effort when the necessary precautions are not taken. Rapid thawing is needed to prevent ice recrystallization in which small crystals grow to sizes that damage cells. Thawing can be performed by incubation for 1 to 2 minutes in water baths at 40°C, and the culture medium should be added gradually to dilute the cryoprotectant substances.

It has been well documented that in addition to vitrification, at least two key factors, dehydration and the formation of ice crystals, affect the survival rate of cryopreserved plants during slow cooling. Skyba et al. (2011) investigated the effect of these factors in the cryopreservation of *Hypericum perforatum* L. sprouts. The data demonstrated a negative correlation between the cooling rate and recovery after cryopreservation, with a maximum of 34.4% at 0.3°C/min. Dehydration decreases the water content in plant cells and tissues as does the time of exposure to cryoprotectant solutions and/or the increase in the water reduction rate, even at lower cooling rates. Thermal analysis measurements have confirmed a positive correlation between the "slope" of the thermal gradient and the cooling rate.

REFERENCES

Ahloowalia, B.S., Maluszynski, M., Nichterlein, K., 2004. Global impact of mutation-derived varieties. Euphytica 135 (2), 187–204.

Aitken-Christie, J., Kozai, T., Smith, M.A.L., 1994. Glossary. In: Aitken-Christie, J., Kozai, T., Smith, M.A.L. (Eds.), Automation and Environmental Control in Plant Tissue Culture. Kluwer, Dordrecht, pp. 9–12.

Alves, S.A.O., Lemos, O.F., Santos Filho, B.G., Silva, A.L.L., 2011. In vitro protocol optimization for development of interspecific hybrids of oil palm (Elaeis oleifera (H.B.K) Cortés × Elaeis guineensis Jacq.). Journal of Biotechnology and Biodiversity 2 (3), 1–6.

Bairu, M.W., Fennell, C.W., Van Staden, J., 2006. The effect of plant growth regulators on somaclonal variation in Cavendish banana (*Musa* AAA cv. "Zelig"). Sci Hortic (Amsterdam) 108, 347–351.

Balch, E.P.M., Reyes, M.E.P., Carrillo, M.L.R., 2012. *In vitro* conservation of *Turbinicarpus* (Cactaceae) under slow growth conditions. Haseltonia 17, 51–57.

Benson, E.E., Lync, P.T., Stacey, G.N., 1998. Advantages in plant cryopreservation technology: current applications in crop plant biotechnology. AgBiotech News and Information 10, 133–141.

Borém, A., Miranda, G.V., 2009. Melhoramento de plantas [Plant breeding], fifth ed. UFV, Viçosa, MG. 529p.

Cançado, G.M.A., Ribeiro, A.P., Piñeros, M.A., Miyata, L.Y., Alvarenga, A.A., Villa, F., et al., 2009. Evaluation of aluminium tolerance in grapevine rootstocks. Vitis 48 (4), 167–173.

Costa, M.A.P.C., Mendes, B.M.J., Mourão Filho, F.A.A., 2003. Somatic hybridisation for improvement of citrus rootstock: production of five new combinations with potential for improved disease resistance. Australian Journal of Experimental Agriculture 43, 1151–1156.

Ferreira, F.R., 2011. Germplasm of fruit crops. Revista Brasileira de Fruticultura 33 (1), 1–6.

Ferreira, E.A., Pasqual, M., Neto, A.T., 2009. In vitro sensitivity of fig plantlets to gamma rays [Sensitividade in vitro de brotações de figueira à radiação gama]. Scientia Agricola 66, 540–542.

Geerts, P., Druart, P., Ochatt, S., Baudoin, J.P., 2008. Protoplast fusion technology for somatic hybridisation in Phaseolus. Biotechnology, Agronomy, Society and Environment 12 (1), 41–46.

Germanà, M.A., 2011. Gametic embryogenesis and haploid technology as valuable support to plant breeding. Plant Cell Reports 30, 839–857.

Hossain, M.A., Konisho, K., Minami, M., Nemoto, K., Hung, C.D., Trueman, S.J., 2012. Alginate encapsulation of shoot tips and nodal segments for short-term storage and distribution of the eucalypt Corymbia torelliana × C. citriodora. Acta Physiology Plantarum 34, 117–128.

Hung, C.D., Trueman, S.J., 2012. Cytokinin concentrations for optimal micropropagation of Corymbia torelliana × C. citriodora. Australian Forestry 75, 233–237.

Kitto, S.L., Janick, J., 1982. Polyox as an artificial seed coat for asexual embryos. HortScience 17, 488.

Larkin, P.J., Scowcroft, W.R., 1981. Somaclonal variation – a novel source of variability from cell cultures for plant improvement. Theoret. Appl. Genet. 60, 197–214.

Ledo, A.S., Gomes, K.K.P., Barboza, S.B.S.C., Vieira, G.S.S., Tupinambá, E.A., Aragão, W.M., 2007. Cultivo in vitro de embriões zigóticos e aclimatação de plântulas de coqueiro-anão [In vitro culture of zygotic embryos and acclimatization of green dwarf coconut palm]. Pesquisa Agropecuária Brasileira 42 (2), 147–154.

López, C.M.R., Wetten, A.C., Wilkinson, M.J., 2010. Progressive erosion of genetic and epigenetic variation in callus-derived cocoa (Theobroma cacao) plants. New Phytologist Trust 186 (4), 856–868.

Moosikapala, L., Te-Chato, S., 2010. Application of in vitro conservation in Vetiveria zizanioides Nash. Journal of Agricultural Technology, 6 (2), 401–407.

Murashige, T., 1978. The impact of plant tissue culture on agriculture. In: Thorpe, T.A. (Ed.), Frontiers of Plant Tissue Culture. University of Calgary, Printing Services, Calgary. pp. 15–26, 518–524.

Patel, D., Power, J.B., Anthony, P., Badakshi, F., Harrison, J.S.H., Davey, M.R., 2011. Somatic hybrid plants of Nicotiana × sanderae (+) N. debneyi with fungal resistance to Peronospora tabacina. Annals of Botany 108, 809–819.

Pech-Aké, A., Maust, B., Orozco-Segovia, A., Oropeza, C., Klimaszewska, K., 2007. The effect of gibberellic acid on the in vitro germination of coconut zygotic embryos and their conversion into plantlets. In Vitro Cellular and Developmental Biology Plant 43, 247–253.

Pereira, J.E.S., Maciel, T.M.S., Costa, F.H.S., Pereira, M.A.A., 2006. Germinação in vitro de embriões zigóticos de murmuru (Astrocaryum ulei) [In vitro germination of "Murmuru" zygotic embryos (Astrocaryum ulei)]. Ciência e Agrotecnologia 30 (2), 251–256.

Rai, M.K., Asthana, P., Singh, S.K., Jaiswal, V.S., Jaiswal, U., 2009. The encapsulation technology in fruit plants – a review. Biotechnology Advances 27, 671–679.

Razdan, M.K., 2003. Introduction to Plant Tissue Culture. Science Publishers. 375p.

Redenbaugh, K., Paasch, B.D., Nichol, J.W., Kossler, M.E., Viss, P.R., Walker, K.A., 1986. Somatic seeds: encapsulation of asexual plant embryos. Natural Biotechnology 4, 797–801.

Rodrigues, F.A., Soares, J.D.R., Santos, R.R., Pasqual, M., Silva, S.O., 2011. Colchicine and amiprophos-methyl (APM) in polyploidy induction in banana plant. African Journal of Biotechnology, 10 (62), 13476–13481.

Santos, T.C., Arrigoni-Blank, M.F., Blank, A.F., Menezes, M.M.L.A., 2012. In vitro conservation of vetiver accessions, Chrysopogon zizanioides (L.) Roberty (Poaceae). Bioscience Journal 28 (6), 963–970.

Sarasan, V., Ramsay, M.M., Roberts, A.V., 2002. In vitro germination and induction of direct somatic embryogenesis in "Bottle Palm" (Hyophorbe lagenicaulis L. Bailey H. E. Moore), a critically endangered Mauritian palm. Plant Cell Reports 20, 1107–1111.

Sarmah, D.K., Borthakur, M., Borua, P.K., 2010. Artificial seed production from encapsulated PLBs regenerated from leaf base of Vanda coerulea Grifft. ex. Lindl. – an endangered orchid. Current Science 98, 686–690.

Sato, M., Hosokawa, M., Doi, M., 2011. Somaclonal variation is induced de novo via the tissue culture process: a study quantifying mutated cells in *Saintpaulia*. PLoS One 6, e23541.

Shiba, T., Mii, M., 2005. Plant regeneration from mesophyll- and cell suspension-derived protoplasts of *Dianthus acicularis* and characterization of regenerated plants. In Vitro Cellular & Developmental Biology Plant 41, 794–800.

Silva, V.A., Gonçalves, G.F., Pereira, M.S.V., Gomes, I.F., Freitas, A.F.R., Diniz, M.F.F.M., et al., 2013. Assessment of mutagenic, antimutagenic and genotoxicity effects of Mimosa tenuiflora. Brazilian Journal of Pharmacognosy 23, 329–334.

Singh, S.K., Rai, M.K., Asthana, P., Pandey, S., Jaiswal, V.S., Jaiswal, U., 2009. Plant regeneration from alginate-encapsulated shoot tips of *Spilanthes acmella* (L.) Murr., a medicinally important and herbal pesticidal plant species. Acta Physiologiae Plantarum 31, 649–653.

Singh, S.K., Rai, M.K., Asthana, P., Sahoo, L., 2010. Alginate-encapsulation of nodal segments for propagation, short-term conservation and germplasm exchange and distribution of *Eclipta alba* (L.). Acta Physiologiae Plantarum 32, 607–610.

Skyba, M., Faltus, M., Záme, J., Cellárová, E., 2011. Thermal analysis of cryopreserved *Hypericum perforatum* L. shoot tips: cooling regime dependent dehydration and ice growth. Thermochimica Acta 514, 22–27.

Soares, J.D.R., Rodrigues, F.A., Pasqual, M., Nunes, C.F., Araujo, A.G., 2011. Germinação de embriões e crescimento inicial in vitro de macaúba [Germination and early growth of embryos of macaúba seedlings]. Ciência Rural, Santa Maria 41 (5), 773–778.

Souza, A.S., Duarte, F.V., Santos-Serejo, J.A., Junghans, T.G., Paz, O.P., Montarroyos, A.V.V., et al., 2009. Preservação de germoplasma vegetal, com ênfase na conservação in vitro de variedade de mandioca [Preservation of plant germplasm with emphasis on in vitro conservation of cassava variety]. EMBRAPA, Cruz das Almas, 24p (Technical Newsletter 90).

Uozumi, N., Nakashimada, Y., Kato, Y., Kobayashi, T., 1992. Production of artificial seed from horseradish hairy root. Journal of Fermentation and Bioengineering 74, 21–26.

Withers, L.A., 1985. Cryopreservation of cultured plant cells and protoplasts. In: Kartha, K.K. (Ed.), Cryopreservation of Plant Cells and Organs. CRC Press, Boca Raton, Florida, pp. 243–267.

Chapter 8

Transgenic Plants

Francisco Murilo Zerbini, Fábio Nascimento da Silva, Gloria Patricia Castillo Urquiza, and Marcos Fernando Basso
Federal University of Viçosa, Viçosa, Brazil

INTRODUCTION

Humankind has genetically manipulated plant crops for many years through conventional breeding. Until recently, breeding was the only way to introduce phenotypic characteristics of interest, which are determined by genes, to an individual plant or species. The desired phenotypic traits are transferred to the progeny through breeding and selection. However, such conventional methods of genetic manipulation have some limitations, including the sexual barrier between species, phylogenetic isolation barriers between and within genetic groups, a reduced gene pool, and linkage drag. All of these drawbacks are in addition to the lengthy time usually required for desirable traits to be transferred.

Recently, molecular biology, tissue culture, and gene transfer techniques have been combined into a powerful tool for the introduction of new traits into a particular plant. As a result, scientists can introduce genes from animals, microorganisms, or different plant species into the genome of a recipient plant in a controlled manner, independent of fertilization. Hence, the sexual barriers between species are eliminated, as well as the barriers between organisms of the eubacteria, archaea, and eukarya phylogenetic domains.

Plant genetic transformation can be defined as the controlled introduction of specific segments of DNA into a recipient plant genome, excluding an introduction by fertilization or hybridization. Other non sexual gene transfer techniques exist for plants, such as somatic hybridization, in which two individual protoplasts are fused, mixing the nuclear and cytoplasmic genomes. Another non sexual transfer method is cybridization, which occurs through cytoplasmic fusion (Lindsey, 1992). However, genetic transformation is the only technique that allows both the insertion and integration of specific DNA fragments.

To insert a desired gene into the host, extensive research in basic molecular and cellular biology is necessary to locate and isolate the gene responsible for the trait of interest in the donor organism (Brasileiro and Carneiro, 1998).

Biotechnology Applied to Plant Breeding. http://dx.doi.org/10.1016/B978-0-12-418672-9.00008-8

Genetic transformation in plants was only made possible by the development of plant tissue culture techniques. For the transformation process to be effective, foreign DNA must first be introduced into a plant cell or plant tissue; a transgenic plant is then regenerated from the transformed cell. This process is enabled by the totipotency phenomenon, which is the ability of individual plant cells to develop into entire new plants under favorable conditions and in the presence of growth regulators and nutrients.

ORGANIZATION AND GENE EXPRESSION IN EUKARYOTES

An understanding of gene organization and gene expression is essential for genetic engineering processes; therefore, a brief overview of these topics, as well as the methodologies used for the manipulation of nucleic acids, is presented.

A gene may be defined as a sequence of nucleotides that is necessary and sufficient for the synthesis of a polypeptide or a stable RNA molecule such as a ribosomal or transfer RNA (rRNA or tRNA). According to this definition, each gene has a coding region consisting of a nucleotide sequence that encodes the amino acid sequence of a polypeptide chain or stable RNA, as well as regulatory nucleotide sequences that determine and control the gene's transcription. Important regions for transcription control are the promoter and terminator sequences. The DNA promoter region is the location at which the enzyme responsible for transcribing the gene, RNA polymerase, binds. The terminator region, on the other hand, is the nucleotide sequence that determines the detachment of RNA polymerase from the DNA template strand, which occurs towards the end of the transcription process. However, in eukaryotes, other regulatory sequences can be present both upstream and downstream of the gene.

RNA polymerase reads the genetic code of a DNA strand and produces a nucleotide sequence called messenger RNA (mRNA), which contains the information for assembling a peptide chain. Gene expression as a process includes both a gene's transcription (the synthesis of a functional mRNA from a DNA nucleotide sequence) and the eventual translation of the mRNA into a corresponding sequence of amino acids. Genes encoding rRNAs, tRNAs, or other smaller classes of RNAs are transcribed but never translated, which only occurs with mRNA that is synthesized from genes encoding polypeptide chains. The process of RNA synthesis, also known as transcription, proceeds in the direction from the 5′ untranslated region (5′UTR) to the 3′ untranslated region (3′UTR), which determines the orientation of the genetic code. In eukaryotic genes, coding regions have nucleotide sequences called introns inserted between them; introns interrupt the coding region. Introns that are present in genomic DNA are transcribed into precursor RNA (pre-mRNA) and removed during RNA splicing, giving rise to mature mRNA. The sequences that are not removed during processing are known as exons. Mature mRNAs are made of sets of three

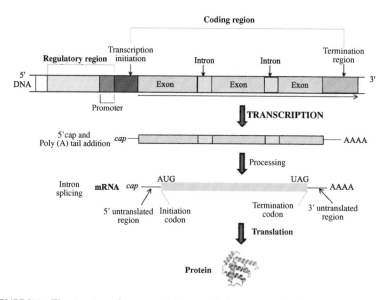

FIGURE 8.1 **The structure of a gene and the control of gene expression in eukaryotes.** Initially, RNA polymerase binds to the promoter region and reads the DNA from the 5′ UTR to the 3′ UTR. A pre-mRNA strand is synthesized and receives a cap (consisting of the modified nucleotide 7-methyl-guanosine triphosphate, or m7GpppNp) to the 5′ end and a repeated adenine sequence (a poly-A tail) to the 3′ end. The pre-mRNA is spliced to remove the introns and transported from the nucleus to the cytoplasm. In the cytoplasm, the mature mRNA is coupled to ribosome subunits. Once bound to the ribosome, the mRNA receives its aminoacyl-tRNA, which contains amino acids specifically matched to each anti-codon, and the polypeptide chain is synthesized.

neighboring nucleotides, called codons, that each code for a specific amino acid. The mRNAs bind to ribosomes to initiate the translation process, in which rRNAs, essential for protein synthesis, read the codons; tRNAs are the molecules responsible for bringing the correct amino acid to add to the chain. Thus, an mRNA is translated into a protein (Figure 8.1).

MANIPULATION OF NUCLEIC ACIDS

As mentioned previously, several steps are involved in the process of creating transgenic plants, including gene identification, isolation, and sequencing; another critical phase is cloning engineering, which involves the construction of the insert containing the known coding region and its promoter and terminator regions. To isolate the genes of interest, a genomic library must be created and probed, and if the identified gene is expressed in a sexually compatible species, conventional breeding can be employed. All of these processes are collectively known as recombinant DNA technology, and they rely on a number of innovative technologies. Some of these technologies are discussed in this chapter, including restriction enzymes, DNA ligases, cloning vectors, bacterial transformation methods and methods to distinguish transformed cells from those that were not transformed.

Restriction Enzymes

Restriction enzymes, also called restriction endonucleases, recognize a specific sequence of nucleotides in double stranded DNA and cut the DNA at a specific location. They are indispensable to the isolation of genes and the construction of cloned DNA molecules. Most restriction enzymes recognize sequences of four to eight base pairs and hydrolyze a single phosphodiester bond on each strand. A characteristic of many of these cleavage or restriction sites is their double rotational symmetry. Generally, the cleavage sites are symmetrically positioned, or palindromic. Restriction enzymes can create fragments with sticky ends, as is the case with the enzyme *Bam*HI, or blunt ends, as with *Hae*III (Table 8.1).

DNA ligases are used to join the fragments of DNA generated by restriction enzymes. The availability of various types of restriction enzymes and ligases enables the transfer of specific DNA sequences from one molecule to another.

Cloning Vectors

Once identified and isolated, the DNA fragment containing the genetic information of interest is inserted into a DNA molecule known as a molecular cloning vector, which is capable of copying and amplifying genetic information. The primary

TABLE 8.1 The Specificity of Some Restriction Enzymes

Name	Origin	Recognition Sequence
*Bam*HI	*Bacillus amyloliquefaciens* H	5′ G//GATCC 3′
		3′ CCTAG//G 5′
*Eco*RI	*Escherichia coli* RY13	5′ G//AATTC 3′
		3″CTTAA//G 5′
*Hind*III	*Haemophilus influenzae* Rd	5′ A//AGCTT 3′
		3′ TTCGA//A 5′
*Hae*III	*Haemophilus aegyptius*	5′ GG//CC 3′
		3′ CC//GG 5′
*Kpn*I	*Klebsiella pneumoniae*	5′ GGTAC//C 3′
		3′ C//CATGG 5′

Double bars indicate the cleavage site in the DNA strand.

cloning vectors used in plant transformation are derived from bacterial plasmids. Plasmids are circular DNA molecules that can be replicated independently of the chromosomal DNA. The recombinant DNA molecule, composed of the plasmid vector and the insert construct, is introduced into a suitable bacterial host cell by a process called transformation. After the recombinant DNA is introduced into the bacterium, the organism generates multiple copies of the DNA of interest, amplifying it. Other cloning vectors used can include bacteriophage DNA, eukaryotic viral DNA and bacterial or yeast artificial chromosomes (BAC or YAC).

All cloning vectors should contain sequences allowing their autonomous replication in the host cell as well as at least one cloning site; specifically, a unique restriction site is needed that is not present in the insert construct and allows for site-specific gene insertion. In addition, the vector must have a selection gene, which is typically a gene that confers a resistance to an antibiotic or other chemical. The inclusion of this gene enables researchers to select the host cells that acquired the vector from those that did not based on their growth in chemical-containing media.

Genomic and cDNA Libraries

Genomic libraries are constructed by cloning the entire genome of an organism. First, a collection of all of a species' DNA fragments is generated using restriction enzymes or mechanical shearing. These fragments are linked to a vector capable of receiving large, lengthy inserts (such as YACs and BACs) so that a smaller number of clones are required to obtain a representative sample of the entire genome. Large inserts also allow for the cloning of a complete gene in a single clone or a small number of different clones, including its 5′ and 3′ flanking sequences.

However, because the genomic DNA is randomly fragmented, only some of the fragments will contain coding genes, and many will still have only a portion of the coding gene. In addition, clones derived from the genomic DNA of a eukaryotic cell will also include non coding sequences (introns).

An alternative strategy to isolate the gene of interest is to build a database containing only the DNA sequences of transcribed mature RNAs, which correspond to functional genes. This strategy can be accomplished by extracting a cell's mRNA molecules and synthesizing from them the complementary DNA (cDNA) that corresponds to each mRNA present. In this method, the mRNA is isolated and purified using the poly-A tail at the 3′ end of eukaryotic mRNAs. The cDNA is synthesized using reverse transcriptase with the mRNA serving as a template. The cDNA is subsequently cloned, and a collection of these clones constitutes a cDNA library.

Once the genes of interest are identified and isolated, they are grouped into an expression cassette, which essentially consists of a promoter, a coding sequence, and a terminator. These cassettes are inserted into a vector suitable for plant transformation, typically one derived from a bacterial plasmid that also contains a transformation tracking marker and a selection gene.

METHODOLOGIES FOR THE DEVELOPMENT OF TRANSGENIC PLANTS

The transfer of exogenous DNA to higher plants can be accomplished by various methods. Here, only the classical and widely used methodologies for the development of transgenic plants will be addressed. Variations to the methods discussed in this chapter as well as other methods for the development of transgenic plants can be found in the literature.

Genetic Transformation Using *Agrobacterium tumefaciens*

Agrobacterium tumefaciens is an aerobic, Gram-negative bacterial species that belongs to the Rhizobiaceae family and is found in soil (Zambryski, 1988). It is a pathogen that infects a wide range of host plants (over 600 species). *Agrobacterium tumefaciens* causes crown gall, a tumor-forming disease, in infected plants, and it has the ability to transfer part of its genetic material into the host plant, altering the gene expression of the host for its own benefit. The related species *A. rhizogenes*, which causes the proliferation of secondary roots at the point of infection, can also be used to transform plants.

During the initial stage of the *A. tumefaciens* infection process, the bacterium will recognize and contact the plant cell. This process of recognition and contact is mediated by signaling molecules such as low molecular weight phenolic compounds, amino acids, and sugars released by the injured root. After contact is established, the signaling molecules promote the activation of virulence genes (*vir* genes) located in the Vir region of the tumor-inducing (Ti) plasmid in *A. tumefaciens* (Figure 8.2). There are six groups of genes in the *vir* region that are either essential (*virA*, *virB*, *virC*, and *virD*) or function to increase the efficiency of the host cell's transformation (*virE* and *virG*). Additionally, other virulence genes located on the bacterial chromosome are involved in the early recognition and contact stages of the infection process (Zambryski, 1988).

The proteins encoded by *vir* genes are essential for the transfer of the T-DNA (transferred DNA) region of the *A. tumefaciens* Ti plasmid to the nucleus of the host plant cell. The T-DNA region is flanked by left and right border segments, which are repeated sequences of approximately 25 base pairs. The left and right borders are critical for the recognition of the transfer region that must be released.

A single strand of the T-DNA complex is transferred, along with the *vir* genes, to the plant cell nucleus, where it is inserted and expressed. It is believed that its integration into the genome is random, but there are indications that integration occurs specifically in transcriptionally active regions (Gelvin, 2003). Regulatory elements present in the T-DNA of *A. tumefaciens* (a prokaryote) allow for transcription of the genes present in this region using the plant eukaryotic transcription system.

The T-DNA of *A. tumefaciens* contains genes encoding enzymes related to the biosynthesis of cytokinins and auxins, which cause the uncontrolled

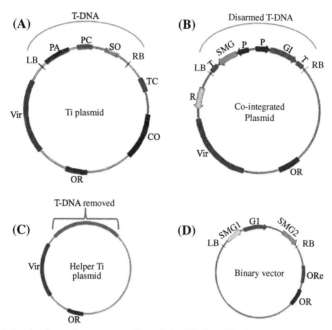

FIGURE 8.2 A schematic representation of the Ti plasmid of *Agrobacterium tumefaciens* and its modifications for plant transformation. (A) The *A. tumefaciens* native Ti plasmid. The oncogenes responsible for the production of cytokinin (PC) and auxin (PA) as well as the gene responsible for opine synthesis (SO) are found in the transferred DNA (T-DNA) region. The T-DNA region is flanked by the Right Border (RB) and Left Border (LB) regions, which are both essential for the T-DNA transfer process. In addition, the Ti plasmid has a virulence region (Vir), a region responsible for conjugative transfer (TC), a region responsible for the expression of genes involved in opine catabolism (CO) and the origin of replication for *A. tumefaciens* (OR). (B) To disarm the Ti plasmid, the PC, PA, SO, TC, and CO regions of the T-DNA are removed, and an exogenous T-DNA section is inserted. The exogenous T-DNA introduces the exogenous gene of interest (GI), the selection marker gene (SMG), and the promoter (P) and terminator (T) regions. Additionally, an antibiotic resistance gene (R) is inserted to allow for selection of the transformed bacteria. (C) The Ti plasmid with excised T-DNA is used as a helper in the T-DNA binary vector system. (D) A binary vector has the GI as well as the SMGs for both *A. tumefaciens* (SMG2) and plants (SMG1). It also has an origin of replication for *E. coli* (Ore). *A. tumefaciens* bacteria carrying a helper Ti plasmid containing the binary vector can be used for plant transformation.

proliferation of transformed cells, leading to tumor formation. In addition, the T-DNA has genes encoding enzymes that are involved in the synthesis of opines (modified amino acids or carbohydrates). An *A. tumefaciens* infection induces the production of large opine quantities, which are catabolized by the bacteria.

A. tumefaciens infections allow portions of a prokaryotic genome (the T-DNA region of Ti plasmid) to be incorporated into the DNA of a eukaryotic organism, a natural property unique to this bacteria. Understanding this natural process of transferring genetic elements between different organisms was essential to obtain transgenic plants.

Clearly, the presence of genes encoding opines and growth regulators is not favorable for the production of transgenic plants – the expression of these genes affects and can prevent the regeneration of the transformed plants. To overcome this problem, the *A. tumefaciens* strains that are used for plant transformations are disarmed by removing the Ti plasmid regions with genes that affect plant development. The removal of these regions is performed by a double homologous recombination procedure that occurs in three stages: (1) the introduction of a secondary plasmid that contains regions of homology with the Ti plasmid of *A. tumefaciens*; (2) a double recombination, in which the undesired regions of the Ti plasmid are integrated into the secondary plasmid, causing the deletion of genes involved in the synthesis of opines and growth regulators; and (3) the recovery of the disarmed Ti-plasmid and the secondary plasmid now containing the deleted region.

After the regions that are detrimental to transformation are removed, a plasmid with the gene of interest can be constructed. Because the Ti plasmid is very large (approximately 200,000 base pairs), it is necessary to use smaller-sized plasmids that are easier to handle. Essentially, these vectors should contain the T-DNA flanking regions, and they may be integrated into a disarmed Ti plasmid to produce a co-integrated plasmid (Figure 8.2B) or simply be a binary vector, independent of a Ti plasmid (Figure 8.2D). The binary vectors are maintained separately from Ti plasmids and are derived from plasmids that can replicate in both *Escherichia coli* and *A. tumefaciens*. Co-integrated vectors result from the integration of intermediary vectors that do not replicate in *A. tumefaciens* with the Ti plasmid, which occurs via simple recombination using a region homologous to the disarmed strain.

In addition to the origin of replication, which is essential for its production in bacteria, the vector for plant transformation must also contain a gene for selection in bacteria. Several selection marker genes can be used, but they all typically confer antibiotic or herbicide resistance to the transformed cells. Also needed are a multiple cloning site between the edges of the T-DNA where the gene of interest can be inserted, promoter and terminator regions for the different genes contained on the plasmid, and elements necessary for conjugation.

Regardless of the type of vector used (binary or co-integrated), the efficiency of the transfer of T-DNA into the plant cell nucleus is similar. Currently, the binary system is the vector most commonly used due to its success in obtaining recombinant strains and the freedom of using any strain of *A. tumefaciens*.

The construction of the vector is followed by its transfer into *A. tumefaciens*. Three common methods are used for vector transfer: (1) triparental conjugation; (2) electroporation; and (3) heat shock. The choice of method depends primarily on the resources available in the laboratory, as the three methods are equally effective.

Triparental conjugation is based on the use of donor, recipient, and helper bacteria. For this method, two strains of *E. coli* (the donor and the helper) and a strain of recipient *A. tumefaciens* are cultured together. The helper *E. coli* promotes the mobilization and transfer of the plasmids between compatible bacteria. This process is mediated by mobilization (*mob*) and transfer (*tra*) genes in the helper's plasmid. The donor *E. coli* carries the binary vector that will be transferred to the recipient bacterium. During the co cultivation of these bacteria, the helper's plasmid is transferred to the donor *E. coli*. Subsequently, this plasmid, now present in the donor bacterium, promotes the mobilization of both itself and the binary vector to *A. tumefaciens* (Brasileiro and Carneiro, 1998). The helper plasmid is unable to replicate in *A. tumefaciens* and is spontaneously eliminated. The strain of recombinant *A. tumefaciens* is then selected as usual with the appropriate antibiotics.

In electroporation, the vector (binary or co-integrated) and *A. tumefaciens* competent cells are together subjected to a high voltage electrical pulse. The generated electric field promotes a reversible destabilization of the plasma membranes, allowing the vector to enter the *A. tumefaciens* bacterium.

In the heat shock method, the permeability of the *A. tumefaciens* plasma membrane is reversibly changed by successive incubation at extreme temperatures (from -186 to $37\,°C$). These changes to the plasma membrane allow the vector into the *A. tumefaciens* cell (Brasileiro and Carneiro, 1998).

The basic principle of plant transformation as mediated by *A. tumefaciens* is the selection of one or more transformed cells and the regeneration of these cells into a transgenic plant. Many different types of plant tissue, including somatic embryos, tubers, cotyledons, and fragments of leaves, stems, or roots, can be used for the co culturing stage, in which both the bacteria and the plant cells are in contact. Phenolic compounds, such as acetosyringone, hydroxyacetosyringone, and others, are added to the co culture medium to facilitate the initial steps of transformation and enhance the recognition, contact, and transfer of the T-DNA. After the co culturing period, the explants (pieces of plant tissue) are maintained in a specific medium containing a selection agent and antibiotics that prevent the regeneration of untransformed *A. tumefaciens*. The dosage and duration of exposure to the selection agent may be adjusted as necessary. After several successive cultures, only the transformed tissues will regenerate and grow into transgenic plants. Leak occurrences are possible, in which non transformed cells can regenerate in the selective medium. Therefore, the transgenic state of the plants must be confirmed before greenhouse acclimation.

Genetic transformation using *A. tumefaciens* is the most commonly used process to obtain transgenic plants; it is low cost and relatively simple compared to other methods. Furthermore, this genetic transformation process results in fewer copies of the transgene and is considered more accurate than direct methods. However, *A. tumefaciens* has a very low capacity to infect monocots, which limits the application of this method to an important group of plants that includes corn, rice, and wheat, among others.

Direct Methods

The introduction of genes into the nucleus of target cells by direct methods is dependent on making physical and chemical changes to the plant's cell walls and membranes. The exogenous gene to be introduced into the cell nucleus is inserted into a vector. The vector is then adhered to a metallic microparticle for its introduction to and transformation of the recipient cell. Theoretically, direct methods may be used to transfer DNA to virtually all plant species.

Particle Bombardment (Biolistics)

For exogenous DNA to enter the nucleus of a host cell, it must overcome the cell wall and the plasma membrane. Particle bombardment is the acceleration of inert metal microparticles that are carrying DNA into the target cell. Of course, these microparticles must overcome the cell's barriers without causing cell death.

Initially, the exogenous DNA must be incorporated into bacterial plasmids. In addition to the exogenous DNA of interest, the vectors should contain a promoter, a terminator, and marker genes for tracking transformation and selection. A high concentration of the DNA vector is used when attaching the vector to the microparticles. In this step, carrier DNA (usually salmon sperm DNA) can also be added to increase the transformation efficiency. This complex of vector, carrier DNA, and metal microparticles must be accelerated to reach the inside of the recipient plant cell nucleus.

Most of the systems used for accelerating the metal microparticles are based on generating great power to displace a membrane that contains the DNA-coated microparticles. Speeds above 1500 km/h are often reached. The energy can be generated from various sources: (1) a chemical explosion (dry gunpowder); (2) a helium gas discharge under high pressure (Figure 8.3); (3) the evaporation of a drop of water through an electrical discharge with either high voltage and low capacitance or low voltage and high capacitance; and (4) the discharge of compressed air. All particle acceleration processes are carried out under a vacuum to minimize the deceleration. When a membrane carrier is not used, the microparticles are accelerated by a helium gas spray discharge at low pressure (Takeuchi et al., 1992). The systems that use helium gas under high pressure and electric discharge are the most efficient at transforming different plant species.

High-density microparticles, such as gold and tungsten, are commonly used in the transformation process. These microparticles measure 0.2 to 3 μm in diameter and should be chemically inert to avoid damage to the exogenous DNA or cellular components. Gold microparticles are more uniform in size and shape and can reach deeper layers of tissue when compared to tungsten. However, the irreversible agglomeration of gold microparticles in aqueous solution can hinder the introduction of exogenous DNA, the only disadvantage of this type of microparticle. Tungsten microparticles can oxidize with time, which may harm the recipient cell and obstruct the adhesion of exogenous DNA.

FIGURE 8.3 A microparticle accelerator chamber, into which the plant material is placed for the biolistic transformation process.

Upon the microparticle's entry into a cell, the attached exogenous DNA dissociates by interaction with cellular fluids and is randomly integrated into the recipient cell genome. Various types of plant tissue, such as callus, cotyledon, leaf discs, embryos, and cell suspensions, have been used in microparticle bombardment. The procedures for the selection and regeneration of transformed cells are essentially the same as those described for genetic transformation via *A. tumefaciens*.

Protoplast Electroporation

Protoplast electroporation is used to introduce exogenous macromolecules such as DNA to plant cells by reversibly changing the permeability of the plasma membrane. Protoplasts are plant cells that have been stripped of their cell walls through the action of pectinases and cellulases. As described for the other plant transformation techniques, it is necessary to obtain a vector containing the exogenous DNA of interest and a selection gene.

To introduce the exogenous DNA into the recipient cell interior, purified protoplasts are mixed with the vectors and the carrier DNA; this mixture is then exposed to short pulses of a continuous current with high voltage. Pores form temporarily in the plasma membrane, allowing the vector carrying the exogenous gene of interest to enter. By increasing the intensity or duration of the pulses, it is possible to increase the number and size of the pores formed (Brasileiro and Carneiro, 1998). The voltage and capacitance intensities, the pulse duration, the electroporation buffer composition, and the concentrations of protoplasts and DNA (both vector and carrier) should be optimized to achieve a successful transformation of protoplasts by electroporation.

The major difficulty with this technique is the regeneration of plants from the transformed protoplasts. Plants regenerated from protoplasts may also have fertility problems. If, when creating the protoplasts, the enzymatic treatment is replaced by a partial plasmolysis of the cells immediately before electroporation, these problems with protoplast regeneration can partially be avoided (Sabri et al., 1996).

As an alternative to electroporation, polyethylene glycol (PEG) associated with calcium and magnesium at an alkaline pH can be used to promote the binding of exogenous DNA to protoplasts. In this method, DNA adheres to the cell's surface and is absorbed by endocytosis. Treatment with PEG can also affect the regeneration of plants from protoplasts, increasing the difficulty of obtaining transformed plants.

LABORATORY STEPS FOR THE DEVELOPMENT OF TRANSGENIC PLANTS

Several specific protocols for plant transformation have been described in the literature. This section will only address the general laboratory steps needed to obtain transgenic plants: (1) the isolation and cloning of the gene of interest; (2) the preparation of the plant material (fragments of leaves, embryos, protoplasts, and others); (3) the DNA transfer (gene transformation); and (4) the regeneration of a plant from the transformed cell.

Isolation and Cloning of the Gene of Interest

In this step, the gene of interest is isolated from the total DNA of the donor organism and cloned into a vector to transform *E. coli* competent cells. The gene of interest can be isolated by constructing a genomic library and using appropriate probes to recognize the gene, as well as by amplifying the gene using the specific primers for the coding region. The isolated gene of interest is used for vector construction. In addition to the transgene, the vector must contain a selection marker gene to track the successfully transformed plant cells. In the case of genetic transformation by *A. tumefaciens*, the gene of interest should be either transferred to a binary vector or co-integrated into the bacteria's Ti plasmid.

Preparation of the Plant Material

This step must be performed under aseptic conditions to avoid contaminating the plant material that will be transformed. The plant tissue to be used should be sterilized with ethanol and sodium hypochlorite.

To produce protoplasts, leaf fragments must be kept in an enzyme solution with the cut surface in contact with the solution. Leaf fragments that have been macerated by enzymatic action are strained, and the recovered solution contains the protoplasts. This solution is subjected to successive centrifugations, and the protoplasts are then resuspended in a specific buffer. Subsequently, the protoplasts are counted in a Neubauer chamber. After being counted, the protoplasts may be used for electroporation or for treatment with PEG.

When using *A. tumefaciens* for genetic transformation, incisions must be made in the explants to facilitate the process of infection and the transfer of exogenous DNA. These cuts are made at the time of preparation, slicing the explants into small pieces. Before their co cultivation with *A. tumefaciens*, the explants are kept in a regeneration medium consisting of full strength Murashige and Skoog (MS) salts (Murashige and Skoog, 1962), vitamins, zeatin, sucrose, and acetosyringone.

DNA Transfer

The procedure of transferring DNA to the target plant cell will depend on the processing method used. An overview of the different procedures is given here. Specific conditions will vary with the plant species and type of tissue used.

In the transformation method mediated by *A. tumefaciens*, its co cultivation with the explants is carried out after the pre culture step. First, the culture of *A. tumefaciens* with the plasmid containing the gene of interest is incubated at 28 °C in a liquid medium with specific antibiotics, and the cultures are then subjected to centrifugation. The bacterial cells are resuspended in liquid MS medium (Murashige and Skoog, 1962) supplemented with myo-inositol and sucrose. An *A. tumefaciens* suspension with an OD_{600} of 0.4 is then made, and the explants are soaked in this suspension to allow the DNA to transfer. Then, the explants are dried with sterile filter paper and transferred back to the pre cultivation medium. The explants are kept in this medium for approximately 2 days before beginning the regeneration step.

In the biolistic method, the particles of tungsten or gold must be sterilized with ethanol, rinsed with sterile distilled water, and resuspended in glycerol. The DNA is precipitated onto the microparticles using calcium chloride and spermidine, and the coated microparticles are subsequently aliquoted onto the carrier membrane. The membrane is placed on a support with a steel retention screen within the bombardment apparatus. Once the plant material has been positioned within the vacuum chamber, the microparticles are accelerated.

In electroporation, the quantified protoplasts are placed in an electroporation cuvette with carrier DNA and plasmids containing the gene of interest, and electrical pulses are applied. The voltage and capacitors used depend on the electroporator and the diameters of the protoplasts. After electroporation, the protoplasts are transferred to a Petri dish containing an appropriate culture medium with auxin and benzylaminopurine. The protoplasts are allowed to rest in this medium and are later transferred to the regeneration medium.

Treatment with PEG can replace electrical pulses as the method of plasmid intake into the cells. The appropriate concentration and molecular weight of the PEG must be established prior to attempting this treatment. In general, 15 to 25% of PEG 6000 is used. After PEG treatment, the protoplasts are transferred to the regeneration medium.

Regenerating Plants from the Transformed Cells

Undifferentiated cells are totipotent, able to differentiate into any type of plant tissue. When explants or protoplasts are placed in contact with a selective regeneration medium, calli (undifferentiated cells) are formed. After this regeneration step, the calli are transferred to an elongation medium for approximately 30 days. The differentiated and elongated shoots are sectioned and transferred to the rooting medium. Rooted plants with developed shoots are then transferred to a substrate in the greenhouse. Once the expression of the transgene is confirmed, the hemizygous transgenic plant (called T_0 or R_0) is self-fertilized to obtain homozygous transgenic progeny.

IDENTIFICATION OF TRANSGENIC PLANTS

Genetic transformation techniques usually transfer the DNA of interest to only a few cells. Thus, to create transgenic plants from this pool of cells it is necessary to use an efficient protocol for regeneration combined with a selection system. For this reason, selection marker genes are typically inserted alongside the gene of interest. The selection genes produce proteins with enzymatic activities that provide some readily detectable characteristic to the transformed cells, such as fluorescence emission or the ability to grow in the presence of a normally toxic substance. These genes are known as marker genes or reporter genes, depending on how the introduced trait is detected.

Marker Genes

Marker genes confer a resistance to toxic substances, such as herbicides and antibiotics, to the plant cells. Detection of the characteristic is performed by adding the toxic compound to the culture medium so that only plant cells expressing the marker gene are able to grow normally.

The most commonly used selection agent in plant transformation is the *nptII* gene, isolated from *E. coli*. This gene encodes neomycin phosphotransferase, which confers resistance to antibiotics such as kanamycin, gentamicin A, neomycin, and others when expressed in plants. The *hpt* gene, also isolated from *E. coli*, codes for hygromycin transferase and confers resistance to the antibiotic hygromycin.

Herbicide resistance genes have also been frequently used as selection markers. One common option is the *bar* gene, which is isolated from the bacterium *Streptomyces hygroscopicus* and encodes phosphinothricin acetyltransferase, an enzyme that inactivates the herbicide phosphinothricin. The *aroA* gene, isolated from the bacterium *Salmonella typhimurium*, is another widely used marker that confers resistance to the herbicide glyphosate.

Reporter Genes

Reporter genes allow for the identification of transformed cells without the need for selective media, as they encode proteins that are more readily apparent. The *gus* gene, which encodes the enzyme β-glucuronidase, is extensively used as a marker of plant transformation. The presence of this gene may be detected by a histochemical evaluation, which forms a blue precipitate in the presence of the appropriate substrate.

Other commonly used reporter genes include the *luc* gene, which encodes the luciferase enzyme from the firefly *Photinus pyralis*, and the *gfp* gene, which encodes a fluorescent protein isolated from the jellyfish *Aequoerea victoria*. Luciferase undergoes a redox reaction to produce yellow-green light that can be detected by a luminometer or scintillation counter. The GFP protein is naturally fluorescent and can be directly detected under ultraviolet light.

The marker gene must be inserted into the plant genome along with the gene of interest. However, recombination events can lead to the insertion of only the marker gene. To ensure that both genes were inserted, plants that are selected by markers or reporter genes should be grown until there is sufficient plant tissue for DNA extraction. This DNA should be tested with PCR or DNA hybridization (Southern blot or dot blot) to confirm the presence of the gene of interest. Transcript detection by Northern blot hybridization and protein detection by either ELISA or Western blot are alternative techniques that can also be used to determine whether the inserted transgene is being expressed.

USE, EFFECTS, AND MANAGEMENT OF TRANSGENIC CULTIVARS

Recombinant DNA technology has enabled the production of plants that express traits outside the species barrier, thus breaking reproductive isolation.

Genetically modified (GM) plants can be classified into three "generations." The first generation was designed to benefit the production process and usually had agronomic traits, such as disease and insect resistance and herbicide tolerance. The second generation was designed to benefit the consumer and includes GM plants whose nutritional characteristics were improved qualitatively and/or quantitatively, as well as plants with better post-harvest quality. The third generation includes GM plants designed to function as biofactories by synthesizing compounds with applications in the pharmaceutical (vaccines, hormones, and antibodies), biosanitation, and manufacturing industries.

Effects on agroecosystems from the use of transgenic cultivars can be direct, due to the presence of functional exogenous genes in the plant, or indirect, resulting from changes in the production management system. These effects can occur both in the soil and in the air. Specifically, changes can occur in the pattern of agrochemical use, conservation practices, agricultural productivity, the ecology of invasive plants, and even the trophic chain (by affecting non target organisms, such as predators, natural enemies, and phytophagous species). Additionally, genes can flow to and cross with wild species, insects and weeds can be selected for resistance, and associated species such as pollinators and herbivorous pests can have altered populations. In the soil, there could be alterations to the organic matter, the composition of root exudates and the mineralization dynamics of chemical elements.

The risk of gene flow in nature is undoubtedly one of the most concerning and controversial direct effects of the cultivation of transgenic plants. When considering the possibility of a vertical gene flow occurrence, or gene transfer from the transgenic cultivar to other non transgenic cultivars of the same species, one must consider that cultivated species are classified into three groups according to their natural rate of self-pollination (allogamous, autogamous, and intermediate). When discussing horizontal gene flow, or gene transfer from the transgenic species to other botanically related species, one must consider that most of the cultivated species in each country were introduced and thus have no related wild species nearby with which they can swap genes.

For allogamous crops such as corn, GM crops should be isolated using minimum insulation distances and physical obstructions. The surviving volunteer plants should be removed from one cycle to the next. If different cultivars are planted, their plantings should occur in flourishing periods that differ by at least 30 days from one area to another. All of these procedures are necessary to contain gene flow and maintain the purity of the seeds produced, avoiding genetic uniformity in the long term (Stewart Jr. et al., 2003).

Other ways to maintain biodiversity and prevent gene flow include the use of exclusion zones (where the planting of a particular GM cultivar is prohibited), the restriction of plantations near protected areas, and the adoption of shelter areas and buffer zones.

The specific legislation on GM crops varies in each country. In the United States, the regulation of GMOs is borne by the US Department of Agriculture (USDA, www.aphis.usda.gov), the Food and Drug Administration (FDA, www.fda.gov), and the Environmental Protection Agency (EPA, www.epa.gov). In Canada, such regulation is under the responsibility of Health Canada (www.hc-sc.gc.ca/index-eng.php), Environment Canada (www.ec.gc.ca), the Canadian Food Inspection Agency (www.inspection.gc.ca), and the Department of Fisheries and Oceans (www.dfo-mpo.gc.ca/index-eng.htm). The performance of biosafety analyses on transgenic varieties falls under the purview of the National Advisory Committee on Agricultural Biotechnology (Comisión Nacional Asesora de Biotecnología Agropecuaria, CONABio) in Argentina and Health Canada in Canada. In Europe, GM regulations are under the European Food Safety Authority (www.efsa.europa.eu), although each country has its own regulatory body. In Australia, the responsible body is the Office for Gene Technology Regulation (www.ogtr.gov.au), while in India, the regulatory agency is Biotech Consortium India Limited. In Brazil, the National Technical Commission on Biosafety (Comissão Técnica Nacional de Biossegurança, CTNBio, www.ctnbio.gov.br) is responsible for biosafety analyses. For example, in the case of corn, CTNBio has outlined coexistence standards that require minimum distances of 100 meters between GM crops and non-GM crops in neighboring areas. However, 20 meters is the minimum as long as a boundary exists of at least 10 rows of conventional plants with sizes and growth cycles similar to the GM corn.

Some strategies for obtaining GM plants can be used to assist in containing gene flow, such as maternal inheritance (using the chloroplast genome), male sterility (intervening in the development of reproductive structures), seed sterility (propagating asexual seed formation, genetic use restriction technology), cleistogamy (causing self-fertilization without flower opening), and genome incompatibility. Additionally, promoters with chemically induced expression that are regulated in a temporal or tissue-specific manner can be used. Mitigation strategies include the use of additional genes that are harmless to GM growth but are deleterious to non target plants in the event of gene flow (Kwit et al., 2011; Daniell, 2002).

For horizontal gene transfer to occur, many natural barriers to the exchange of genetic material must be simultaneously overcome. Therefore, it is assumed that the normal frequency of this phenomenon is extremely low (Kim et al., 2010). The likelihood of genes moving from plants to bacteria involves the availability of target DNA versus competitor DNA, the bacterial cell competence and the integrity of the exogenous eukaryotic DNA, as well as the gene's integration, expression, and selection in the transformed cell. Only in a situation where all of these factors show a positive simultaneous occurrence may gene flow occur.

Insect resistance in GM plants has been one of the key measures used to manage insect pests in various crops. However, due to the constitutive expression of genes that confer resistance to insects and the numerous annual reproductive cycles of these insects, the introduced proteins over time exert great selective pressure on these target pests, enabling the selection of resistant insect populations. Managing GM crops is therefore of fundamental importance to ensure the longevity of the GM variety. Some of the key management strategies of crops with GM insect resistance are the use of refuge areas (to be a source of susceptible insects), highly expressing the insecticidal protein or pyramiding the genes involved. In some cases, however, high protein expression may have an opposite effect, quickening insect resistance, but this can be mitigated by the use of refuge areas and gene pyramiding. In all cases, it is essential to monitor changes in the allele frequencies of the target insects (Tabashnik et al., 2009). However, there are many influences on the food chain and insect biodiversity, such as adaptability and behavior, changes in adult longevity, oviposition effects and the absence of insect hosts or prey. Non target insects may be important agents in the biological control of crop pests, or they may act as pollinators. All of these possible effects should be considered when drawing up a management strategy aimed to minimize imbalances.

Non target insects can acquire the insecticidal protein directly from the sap or pollen of GM plants or indirectly through a prey that consumed the protein. Thus, it is also important to study the route of the insect's intake of this protein (Romeis et al., 2006).

Soil microbes perform several functions that benefit the plant community. The changes in cellular compositions and exudates that accompany the release of insecticidal proteins and antibiotics to the environment can influence the population of these microorganisms, which consequently influences the stability and balance of the agroecosystem. For example, planting GM papaya trees resistant to papaya ring spot virus showed effects on the local soil microbiota, possibly due to the use of a selection gene conferring kanamycin resistance (Wei et al., 2006). On the other hand, planting Bt cotton caused no significant changes to the biochemical properties or microbes of the soil (Shen et al., 2006).

Accordingly, before the commercial release of a transgenic cultivar, it is necessary to develop various studies that evaluate the potential risk of adverse effects on the environment.

Biosafety of Transgenic Cultivars

Biosafety refers to actions taken for the prevention, minimization, and elimination of risks inherent to the research, production, education, technological development, and provision of services related to biotechnology, with an aim to preserve the health of living beings and the integrity of the environment.

Successful biosafety measures are achieved by risk management, which first seeks to identify a potential hazard and assesses the risk involved on a case-by-case basis, using judicious scientific studies. Then, measures are adopted to eliminate or minimize that risk. To assess the biosafety of a GM plant, it is compared to a conventional analog that has a history of safe use. Aspects of this assessment include a molecular characterization of the transgene and the stability of the genetic modification (transcriptome, proteome, and metabolome), proof that no unintended changes occurred at the metabolic level (nutritional equivalence), and toxicological and allergenic profiles indicating no adverse effects. Additionally, the potential vertical and horizontal gene flows must be studied, the possible effects on non target organisms (animals, microorganisms, insects, and humans) must be determined, and the environmental risks must be analyzed.

Brazil has one of the most complete and advanced set of federal laws related to biosafety. In 1995, the first Biosafety Law was enacted, and the CTNBio was created. The CTNBio is composed of regular members and substitute members appointed by the Minister of State for Science and Technology. It advises the President of the Republic in the formulation and implementation of national biosafety laws and establishes technical standards for biosafety, providing conclusive technical advice relating to the protection of human health, living organisms, and the environment in relation to GMOs.

The CTNBio analyzes requests that are submitted on a case-by-case basis, and it is up to the applicant to demonstrate the biosafety of a GMO and provide all the necessary data for its evaluation. According to the procedures specified in the Biosafety Law, the request to release GMOs into the environment is delegated to the specific Health, Plant, Animal, or Environmental Subcommittee, which determines the criteria and makes recommendations for the GMO's release or rejection.

The conclusive technical assessment issued by the CTNBio necessarily includes the following three aspects of GMO biosafety: (1) risks to human health and to food production aimed at human consumption; (2) risks from the agricultural and animal points of view; and (3) risks to the environment. The binding nature of their technical assessment of the environmental and food safety of transgenic organisms widely empowers the CTNBio to monitor the development and scientific technical progress in biosafety and related areas.

The CTNBio classifies the risks associated with GMOs and sets biosafety levels that are to be applied to all activities and projects involving GMOs and their derivatives. GMOs are classified according to four risk categories: the pathogenic potential of the donor and recipient organisms, the transferred nucleotide sequence, its expression in the recipient organism, and the resulting GMO's potential effects on the health of humans, animals, plants, and the environment. Four biosafety levels (BSL-1, BSL-2, BSL-3, and BSL-4) have increasing degrees of containment complexity and protection. Activities and building facilities should be planned and implemented in accordance with the GMO risk classification and biosafety level to prevent an accidental release.

Based on the Biosafety Law, the CTNBio can terminate any activities involving GMOs once the existence of significant risks to human, animals, plants, or the environment has been verified.

To ensure that the released commercial product is continuously monitored, the CTNBio requires that the company or institution holding the rights to cultivate the GMO provide analyses and studies of the commercial plantations for a defined period, even if the initial risk analysis did not indicate safety concerns. Thus, along with the application for commercial release sent to the CTNBio, the applicant must submit a plan for post-release monitoring. The company then has 30 days to implement its proposal after a technical decision favorable to the GMO's commercial use is published. The monitoring includes aspects related to human, animal, and environmental health. For example, the diversity and incidence of weeds in the area is evaluated as a potential result of gene flow; a periodic evaluation must also be performed to determine the fluctuation and population dynamics of the soil's microbes, insects, and pathogens. At the end of each cropping season, an annual report must be sent to the CTNBio, and monitoring areas may undergo scientific audits by expert agencies in the presence of inspectors from the Ministry of Agriculture. The preparation, submission, and subsequent implementation of the overall monitoring plan and/or case-specific post-commercial release plan is the responsibility of the applicant, who can contract out the process to qualified institutions.

If there is an incidence of non compliance with the biosafety regulations or an observed adverse effect caused by the product, the CTNBio can remove the product from the National Registry of Cultivars and therefore from the seed market and the food industry. This removal would occur through the action of supervisory organizations and the Ministry of Agriculture. Thus, the embargo of experiments and plantations with GM cultivars is the prerogative of the supervisory agencies of the Ministries of Health, Agriculture, and Environment.

Commercial Transgenic Cultivars

The commercial use of GM crops is one of the great biotechnology milestones in recent years. The first transgenic plants were generated in the early 1980s, and they have been commercialized since 1994. Since GM's adoption, the global area of GM crops has increased from 1.7 million hectares in 1996 to approximately 170 million hectares in 2012, a 100-fold increase. Currently, 28 countries have adopted the planting of GM crops, including 20 developing countries and eight industrialized countries (James, 2012).

In 2012, the United States planted 69.5 million hectares of GM corn, soybeans, cotton, canola, sugar beets, alfalfa, pumpkin, and papaya. Brazil and Argentina planted 36.6 and 23.9 million hectares, respectively, of GM soybeans, corn, and cotton. These three countries accounted for 76% of the global area planted with GM crops.

Worldwide, soybeans are the primary GM crop, occupying approximately 75.4 million hectares, or 47% of the global GM crop area. Second is corn (51 million hectares, 32%), followed by cotton (24.2 million acres, 15%) and canola (8.2 million hectares, 5%). The predominant agronomic traits introduced into GM plants are herbicide tolerance (59% of the total GM planted area), insect resistance (15%), and the combination of these two characteristics (26%) (James, 2012).

In Brazil, Aragão and Faria (2009) genetically transformed plants using a fragment of a non translatable gene that encodes the viral replication initiator protein (rep or AC1) for RNA interference (RNAi), obtaining a common bean (*Phaseolus vulgaris*) that was resistant to bean golden mosaic virus (*Begomovirus* genus). The transgenic lines expressing the transgene-derived RNA, which was not translatable and therefore would not lead to the production of a protein, showed delayed and attenuated symptoms of golden mosaic due to the activation of post-transcriptional gene silencing. After approximately 10 years of scientific studies, the authors, along with their respective research groups representing Embrapa Rice and Beans (Embrapa Arroz e Feijão) and Embrapa Genetic Resources and Biotechnology (Embrapa Recursos Genéticos e Biotecnologia), applied for a commercial release from the CTNBio and offered their post-commercial release monitoring plan. The CTNBio granted the commercial release after concluding that the GM beans (called Embrapa 5.1) are essentially equivalent to conventional beans and are safe for human and animal use. The GM beans were not considered a potential cause of significant environmental degradation because their environmental studies produced results that were identical to those of conventional beans. This GM plant was the first case of a GM cultivar produced entirely in Brazil by researchers associated with a public company being released commercially. It was also the first Brazilian case of a commercially released GM cultivar with a feature other than herbicide tolerance or insect resistance. Although there are other cases of virus-resistant GM crops in other countries, these beans were the first crop with resistance to the *Begomovirus* genus, one of the most economically important viruses. The commercial release of the GM bean Embrapa 5.1 was a milestone for biotechnology in Brazil.

REFERENCES

Aragão, F.J., Faria, J.C., 2009. First transgenic geminivirus-resistant plant in the field. Nature Biotechnology 27, 1086–1088.

Brasileiro, A.C.M., Carneiro, V.T.C. (Eds.) 1998. Manual de Transformação Genética de Plantas. [Manual of Genetic Transformation in Plants]. Brasília, DF: Embrapa Cenargen. p. 176.

Daniell, H., 2002. Molecular strategies for gene containment in transgenic crops. Nature Biotechnology 20, 581–587.

Gelvin, S.B., 2003. *Agrobacterium*-mediated plant transformation: the biology behind the gene-jockeying tool. Microbiology and Molecular Biology Reviews 67, 16–37.

James, C., 2012. Global status of commercialized biotech/GM Crops: 2012. Available at: http://www.isaaa.org. Accessed on: 29/04/2013.

Kim, S.E., Moon, J.S., Kim, J.K., Choi, W.S., Lee, S.H., Kim, S.U., 2010. Investigation of possible horizontal gene transfer from transgenic rice to soil microorganisms in paddy rice field. Journal of Microbiology and Biotechnology 20, 187–192.

Kwit, C., Moon, H.S., Warwick, S.I., Stewart, C.N., 2011. Transgene introgression in crop relatives: molecular evidence and mitigation strategies. Trends in Biotechnology 29, 284–293.

Lindsey, K., 1992. Genetic manipulation of crops plants. Journal of Biotechnology 26, 1–28.

Murashige, T., Skoog, F., 1962. A revised medium for rapid growth and biossays with tobacco tissue cultures. Plant Physiology 15, 473–497.

Romeis, J., Meissle, M., Bigler, F., 2006. Transgenic crops expressing *Bacillus thuringiensis* toxins and biological control. Nature Biotechnology 24, 63–71.

Sabri, N., Pelissier, B., Teissie, J., 1996. Transient and stable electrotransformation of intact black Mexican sweet maize cells is obtained after plasmolysis. Plant Cell Reports 15, 924–928.

Shen, R.F., Cai, H., Going, W.H., 2006. Transgenic Bt cotton has no apparent effect on enzymatic activities or functional diversity of microbial communities in rhizosphere soil. Plant and Soil 285, 149–159.

Stewart Jr., C.N., Halfhill, M.D., Warwick, S.I., 2003. Transgene introgression from genetically modified crops to their wild relatives. Nature Reviews Genetics 4, 806–817.

Tabashnik, B.E., Rensburg, V.J.B.J., Carriere, Y., 2009. Field-evolved insect resistance to Bt crops: definition, theory and data. Journal of Economic Entomology 102, 2011–2025.

Takeuchi, Y., Dotson, M., Keen, N.T., 1992. Plant transformation: a simple particle bombardment device based on flowing helium. Plant Molecular Biology 18, 835–839.

Wei, X.D., Zou, H.L., Chu, L.M., Liao, B., Ye, C.M., Lan, C.Y., 2006. Field released transgenic papaya affects microbial communities and enzyme activities in soil. Plant and Soil 285, 347–358.

Zambryski, P., 1988. Basic processes underlying *Agrobacterium*-mediated DNA transfer to plant cells. Annual Review of Genetics 22, 1–30.

Chapter 9

Double Haploids

Roberto Fritsche-Neto,[a] Deoclecio Domingos Garbuglio,[b] and Aluízio Borém[c]

[a]Universidade of São Paulo, São Paulo, Brazil, [b]Instituto Agronômico do Paraná, Londrina, Brazil, [c]Universidade Federal de Viçosa, Viçosa, Brazil

Haploids are individuals possessing a single set of chromosomes of the species (n), which is the number of chromosomes in a gamete. Therefore, a haploid plant is considered in the sporofitic stage; however, it has the gametic number of chromosomes typical of its species (Figure 9.1).

The first reports on haploids are from the 1920s. According to Dunwell (2010), the first haploid angiosperm species identified was a dwarf cotton plant. The next reports in the scientific literature were with *Datura stramonium* (Blakelsee, 1922), *Nicotiana tabacum* (Clausen and Mann, 1924), and *Triticum compactum humboldtii* (Gaines and Aase, 1926).

Today, there are several methods to obtain haploids and double haploids (DH), such as inducing-haploid gene, anther and microspore culture and interespecific crosses, among others.

Haploids have little value in agriculture, since haploid plants in general have little vigor, are dwarf, and do not yield much when compared with their counterparts' diploids. They are also more sensitive to biotic and abiotic stresses and show a high degree of sterility.

In plant breeding, haploids also do not have a direct application, but they can be integrated into breeding programs after chromosome duplication, producing the DH. The DH are completely homozygous (all loci are in homozygosity) and the time to obtain them in general is very short, when compared to the many self-pollination generations involved in producing homozygous lines in traditional breeding schemes.

When selfing pollination is used to obtain homozygous individuals, in general five to eight generations are used (Table 9.1). After this, breeders start the agronomy evaluation of the inbred lines.

Especially in self-pollinating crops some breeders start the inbred lines' agronomic evaluation as early as F_4 or F_5, that is, when homozygosity is between 93.75 and 96.88 (Table 9.1) to reduce the time to release a new cultivar. Even when the breeder delays the agronomic evaluation to F_7 the average level of

Biotechnology Applied to Plant Breeding. http://dx.doi.org/10.1016/B978-0-12-418672-9.00009-X

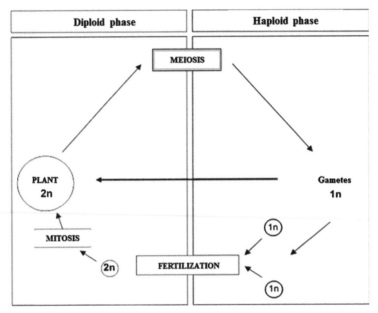

FIGURE 9.1 The different stages of the plant cycle in function of the ploid number.

TABLE 9.1 Percentage of Homozygosity and Number of Self Pollinations Needed to Obtain Inbred Lines in a Traditional Breeding Program of Autogamous Species

	Traditional Method			
Seasons	Generation Planting	Harvesting	% Homozygosity	⊗[1]
1	Crossing block	F1	0%	
2	F1	F2	50%	1
3	F2	F3	75%	2
4	F3	F4	87.50%	3
5	F4	F5	93.75%	4
6	F5	F6	96.88%	5
7	F6	F7	98.44%	6
8	F7	F8	99.22%	7

[1]Number of self pollinations.

homozygosity is 99.22%. Thus, if there were no winter nursery or other strategy to shorten the breeding cycles it would take around to 10 years to develop and launch a new cultivar.

With haploid-inducing genes, the most used technique in the maize breeding programs (Borém and Miranda, 2013), it would take 1 year to obtain the proper population through crosses with the haploid-inducing parent. In this season, breeders select kernels that would generate haploids and chromosome doubling is processed. In the following season, the individuals are grown to generate double-haploid seeds. In the third season, those inbred lines are field evaluated. The more promising lines are selected to enter the yield tests, in the case of autogamous species, or in the test crosses, in the case of alogamous species.

Basically, it takes about to 6 years from crosses to cultivar release, a much faster scheme than in traditional breeding programs, that is, 4 to 6 years quicker.

To be competitive in releasing superior cultivars, breeding programs need to generate a high number of inbred lines or hybrids in a progressively shorter time. The large maize breeding companies have been adopting the inducing-haploid genes to stay competitive in the seed market.

HAPLOID PRODUCTION

A breeding program based on double-haploid lines begins, as with any traditional breeding program, with parent selection. After crosses are done, F_1 plants are obtained and advanced to the F_2 generation when segregation starts, and then the F_2 plants are used as a source of variability to produce the double-haploid lines.

Depending on the used method, F_1 plants or a sample of the F_2 population are crossed with other species or genera (wide crosses), or, as in the case of maize, are crossed with haploid inducing lines. The F_1 plants can be used as the source of ovules, anthers, or microspores, for the production of haploids, via tissue culture (Chen et al., 2011). Although the haploids can be produced from plants in more advanced generations, the gametes from F_1 plants are a good sample of genetic variability that would express in F_2.

A criticism about the use of double-haploid lines obtained from F_1 plants is related to the limited number of opportunities for close genetic linkages to break due to the limited number crossing over. Additionally, it is considered that this potential genetic variability produced by crossing over cannot be observed widely, because there are few generations of recombination and opportunities to observe it (Hu, 1986). Simulation studies by Yonezawa et al. (1987) have shown that haploids from F_2 plants may be more efficient due to the greater segregation and opportunity of linkage breakages with the self-pollination from F_1 to F_2. Hu (1997) and Ma et al. (1999) recommend, as a strategy, that the haploid production starts in F_2 or even in F_3, to allow greater recombination and also the possibility to select for traits of interest. However, the delay to F_3 may be too costly in an era when most breeding programs are trying to be the first to get their varieties to market.

There are several methods for haploid production and the efficiency of each one varies with the species. For example, the method of anther culture is very efficient in barley, wheat, and brassicas, but still has some drawbacks in soybean, common beans, and oats. In maize, the method with better results and the one mostly used by the breeding industry is the use of haploid-inducing lines (Borém and Miranda, 2013).

Tissue Culture

Cellular totipotency was first hypothesized in the mid-19th century, based on the observations of the high capacity of regeneration of plants. In 1953, Muir, cited by Henshaw et al. (1982), was able to regenerate plants from isolated cells, demonstrating the theory of cellular totipotency. With this principle the first protocols for *in vitro* production of haploids were developed in the 1960s and 1970s, via anther culture (Guha and Maheshwari, 1964, 1966). The first commercial cultivars developed by this means were obtained in canola and barley. Thirty years later, Veilleux (1994) reported several cultivars and special germplasms developed by tissue culture in asparagus, maize, rice, tobacco, and wheat. According to Murovec and Bohanec (2012), the number of species where double-haploid plants had been successfully produced via anther culture was over 250. The references of the protocols for inducing haploids for species of apple, blackberry, carnation, cucumber, plum, kiwifruit, mandarin, melon, onion, pear, petunia, rose, *Nicotiana*, squash, sunflower, sweet cherry, watermelon, etc. may be found at Murovec and Bohanec (2012).

The process by which a gametic cell is diverted from its normal organogenic route to develop a somatic cell, via embryogenesis or organogenesis, is responsible for haploid production (Borém and Miranda, 2013).

There are two routes for regeneration of a whole plant from anther culture: gametic embryogenesis and organogenesis (Figure 9.2). In gametic embryogenesis, the microspore or the immature pollen cultivated in a proper nutritive medium rapidly divides forming embryoids. The embryoids developing in the medium form shoots and roots, according to the growth hormone balance. After additional development the plantlets can be transplanted to a greenhouse

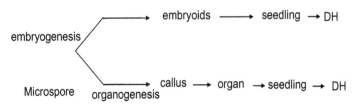

FIGURE 9.2 Embryogenic and organogenic route for double-haploid individual production, via anther culture.

FIGURE 9.3 Somatic embryogenesis in barley: (A) anthers on a media, (B) embryoids, (C) plantlets, and (D) whole regenerated plant. *(Source: Borém and Miranda, 2013)*

for adaptation (Figure 9.3). In the organogenic route the microspores are undifferentiated in an amorphous mass of cells, called callus. This callus can be induced to differentiate into shoots and later also in roots, in a process called organogenesis (Henshaw et al., 1982).

A simplified scheme for double-haploid production via anther culture is illustrated in Figure 9.4.

The complete protocol for double-haploid production via anther culture is described by Bjornstad (1989). This method has been successfully utilized for wheat, barley, and many other species.

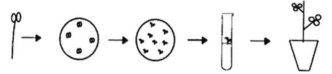

FIGURE 9.4 Double-haploid production via anther culture. *(Source: Borém and Miranda, 2013)*

Interspecific Crosses (Wide Crosses)

Interspecific crosses with the objective of haploid production has been carried out efficaciously in many species of potato, alfalfa, brassica, wheat, oat, triticale, and strawberry, among others (Murovec and Bohanec, 2012).

As in the case of wheat, plants are grown in a greenhouse and their spikelets are emasculated (Figure 9.5A). Three days after emasculation, artificial pollination is carried out using maize pollen (Figure 9.5B). It must be recognized that the origin of the maize pollen affects the frequency of embryo and caryopsis formation. The medium used to germinate the embryos is also critical (Moura et al., 2008). Most programs prefer to perform pollination in the morning, while the air temperature is cool. In general, all spikelets are pollinated again the following day to improve the success rate. 2,4-D and silver nitrate are applied 24 hours after the second pollination, as described by Bidmeshkipour et al. (2007) (Figure 9.5C). By 16 to 18 days after pollination, when the seeds are still immature, the embryos are rescued (Figure 9.5D), and transferred the P2 medium for *in vitro* culture (Figures 9.5E) (Chuang et al., 1978). In this medium the haploid plant can grow up to the time of chromosome duplication (Figure 9.5F).

The development of an individual from unfertilized ovules is called parthenogenesis. Frequently, parthenogenesis occurs *in vivo* and in this case is named polyembriony, as in citrus. Citrus seeds, in general, possess nuclear and zygotic embryos.

After double fertilization, if the endosperm is blocked due to incompatibility of the parent, genomes and the developing seed may abort (Murovec and Bohanec, 2012). However, the rescue of the embryo and its cultivation in a proper medium may allow its development. The function of the abnormal

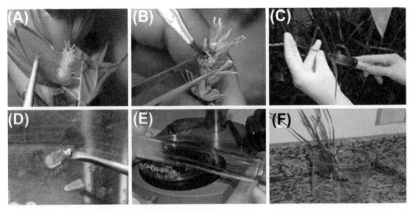

FIGURE 9.5 Wide crosses (interspecific crosses). (A) Emasculation of wheat plants; (B) pollination with maize pollen; (C) application of 2,4-D and AgNO$_3$ solution; (D) embryo rescue; (E) transfer to tissue culture; and (F) double-haploid plant produced.

endosperm is substituted for the tissue culture medium, allowing the complete development of the embryo. Depending on the incompatibility of the parent genomes, diploid hybrids may be formed or, as in many cases, one of the genomes may be eliminated. If the genome of the male parent is completely eliminated a haploid embryo is produced in a process similar to parthenogenesis.

The male parent chromosome elimination process occurs by one of the two ways: (1) mitosis dependent and (2) mitosis independent (Figure 9.6).

In the mitosis-dependent route there is no perfect pairing of chromosomes of the two species during cell divisions, causing asynchrony in the mitotic cycle producing micronuclei. Those micronuclei hold the chromosomes isolated, which would have migrated to the cell poles. As this is not a complete set of chromosomes the actual nucleus recognizes those chromosomes as exotic DNA and causes their heterochromatization and degradation. Therefore, after several mitotic cycles all genetic material from the male parent is eliminated.

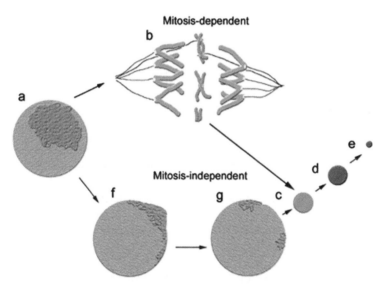

FIGURE 9.6 Schematic for the mitotic and interphase elimination of pearl millet chromo somes from wheat x pearl millet hybrid embryos. Mitotic elimination: (a) spatial separation of parental genomes; (b) imperfect segregation of pearl millet chromosomes caused by (1) faulty kinetochore/spindle fiber interaction, (2) the presence of an additional centromere, or (3) absence of a centromere; (c) formation of micronucleus; (d) heterochromatinization and DNA fragmentation of micronucleus; (e) disintegration of micronucleus. Interphase elimination: (a) spatial separation of parental genomes; (f) budding of pearl millet chromatin; (g) release of pearl millet chromatin-containing micronucleus; (c) formation of micronucleus; (d) heterochromatinization and DNA fragmentation of micronucleus; (e) disintegration of micronucleus. (*Source: Gernand et al., 2005*)

In the mitosis-independent route the cell divisions are not symmetric and there is separation of the genome of each parent. The chromosomes of the male parent migrate to the extremity of the embryo and throughout the cell division they are eliminated.

Haploid-Inducing Lines

In maize, haploids naturally occur at a rate lower than 1:1000 produced seeds (Chase, 1974); however, breeders have been experimenting with different methods to improve this rate. One of them uses haploid-inducing lines. This is a low-cost methodology and also does not depend on special chemicals or expensive equipment (Belicuas et al., 2007).

Most of the haploid-inducing lines are derived from W23 (Kermicle, 1969), which generates androgenetic haploids, and from Stock 6 (Coe, 1959), which produces gymnogenetic haploids. The rate of haploid formation varies from 1 to 2% when using Stock 6 and from 0 to 2% when using W23. However, those rates can vary even more depending on the genetic material from which haploids are being developed (Lashermes and Beckert, 1988).

In recent years new gymnogenetic haploid-inducing lines were developed as the KMS (Tyrnov and Zavalishina, 1984), ZMS (Chalyk, 1994), MHI (Chalyk, 1999), M741H (developed by the Maize Genetics Cooperation-Stock Center, USDA/ARS), RWS (Röber et al., 2005), HZI1 (Zhang et al., 2008), and PHI 1, 2, 3, and 4 (Rotarenco et al., 2010), besides the lines derived from crosses between W23 and Stock 6, such as WS14 (Lashermes et al., 1988). There are reports of frequency of haploid induction surpassing 15% in some materials of private companies, but those results are unpublished.

The mechanisms of *in vivo* haploid induction are not well understood and many hypothesis have been proposed, but most scientists believe the haploid production is due to microsporogenesis abnormalities and/or abnormality in the fertilization process.

Kermicle (1969) observed that the gametofitic indeterminate mutation (ig) in the W23 line is related to some abnormalities in the embrionary sac development, promoting the formation of haploids from male origin. Later, it was observed that ig mutants interfere with the embrionary sac formation, altering the activity of other genes related to this process (Huang and Sheridan, 1996). Hu (1990) observed differences in speed of spermatic nucleus transfer, noting that the one with higher speed has normal fertilization, while the slower one is lost, breaking the double fertilization process producing kernels with haploid embryos. Bylich and Chalik (1996) detected spermatic nucleus morphologically different on the ZMS line, noticing that in one of the nuclei there was normal fertilization, while in the other it was abnormal. Chalyk et al. (2003) reported the occurrence of aneuploidia above 15% in microsporocytes in the MHI haploid-inducing line and only 1% in two lines used as control. They concluded that the abnormality in the microsporogenesis and fertilization may be one of the reasons for haploid induction. Coe and Neuffer (2005) added that the chromosome elimination may be an important mechanism in the *in vivo* haploid induction.

Most breeding programs have focused on the gymnogenetic haploid-inducing lines not only due to their higher rate of haploid induction but also because they are easy work with. In the gymnogenetic system the handling of the field plants is easier since different materials can be induced simultaneously. The genetic materials from which haploids will be induced, that is used as female, should be planted in isolated fields (Figure 9.7) and split into blocks with two to four rows 10 to 20 meters long, alternated with the haploid-inducing line, used as male. Before flowering, the plants are emasculated and wind pollination occurs naturally.

For the production of androgenetic haploids, the individuals from which haploids are to be developed are used as male and the haploid-inducing line is used as female (Figure 9.8). When a single isolated field is used, the crosses must be controlled, requiring intensive labor.

In both mentioned cases the haploid-inducing lines are adapted to a temperate climate. In tropical conditions the haploid-inducing rates may be reduced due to the interaction genotype × environment, but this limitation may be offset with the tropicalization of the haploid-inducing lines (Table 9.2) or even with the use of single-, double-, or three-way cross-inducing hybrids. When using those alternatives one should consider not only the origin of the material to be induced but also the success rate in inducing the haploid formation.

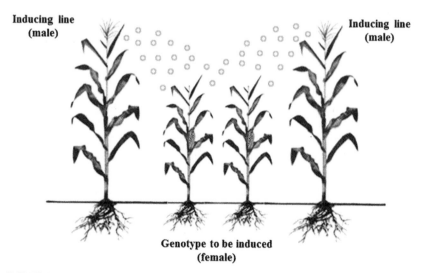

FIGURE 9.7 Haploid production through haploid-inducing lines. (*Source: IAPAR*)

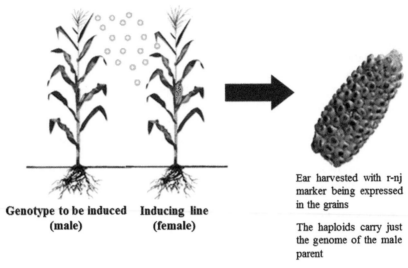

Genotype to be induced   **Inducing line**
(male)   **(female)**

Ear harvested with r-nj marker being expressed in the grains

The haploids carry just the genome of the male parent

FIGURE 9.8   Haploid production through androgenetic inducing lines. (*Source: IAPAR*)

TABLE 9.2 Production of Gimnogenetic Haploid-inducing Stocks, with Different Levels of Tropicalization, Using the Dominant Marker R-navajo (R^{nj}) for Selection

Season	Phase	% Tropical Germplasm
Summer harvest	Temperate inducing line R^{nj} R^{nj} × Tropical line r^{nj} r^{nj}	
	♂ (pigmented seeds) ♀ (no pigmented seeds)	
	Harvested ears: 100% pigmented seeds (R^{nj} r^{nj})	50%
Winter harvest	Plants R^{nj} r^{nj} × Tropical line r^{nj} r^{nj}	
	♂ (pigmented seeds) ♀ (no pigmentation)	
	Harvested ears:	
	50% seed R^{nj} r^{nj} (pigmented seeds)	75%
	50% seeds r^{nj} r^{nj} discarded (no pigmented seeds)	

(Continued)

TABLE 9.2 Continued

Season	Phase	% Tropical Germplasm
Four harvests	(a) May use R^{nj} r^{nj} individuals for self pollinations	75%
	(b) If a higher level of tropicalization is desired four generations of backcrosses should be done	>98%
Summer harvest	Plants R^{nj} r^{nj} (purple embryo): self pollination and harvest	
Winter harvest	Planting and self-pollination of the pigmented seeds to evaluate the following generation	
	Select pigmented ears:	
	33% of ears are R^{nj} R^{nj} (all pigmented seeds)	75%
	66% of ears are R^{nj} r^{nj} (segregate for seed pigmentation)	
Winter harvest	From here on there are two alternatives:	
	(a) Recombination of R^{nj} R^{nj} individuals (33%) to obtain the haploid-inducing population. Beginning of the evaluation of the rate of haploid formation	
	(b) Advance five additional cycles of self pollination of R^{nj} R^{nj} individuals to extract haploid-inducing lines and/or obtain haploid-inducing single-, double-, or three-way cross hybrids.	75%

PLOID IDENTIFICATION AND CHROMOSOME COUNTING

Phenotypical Markers

In maize, for the identification of haploids through *in vivo* methods, an anthocyanin marker is preferably used (Nanda and Chase, 1966). In this system, the *R-navajo* (R^{nj}) gene is used to distinguish haploid from diploid individuals,

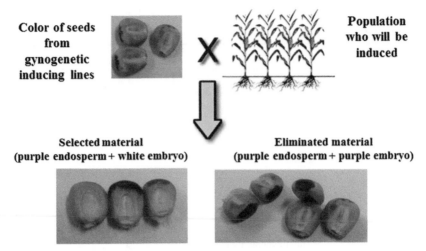

Color of seeds from gynogenetic inducing lines   **X**   **Population who will be induced**

Selected material (purple endosperm + white embryo)   **Eliminated material (purple endosperm + purple embryo)**

FIGURE 9.9   Morphological marker for seed color to identify haploids. (*Source: IAPAR*)

since it promotes pigmentation of the endosperm and the embryo. Purple endosperm seeds with embryo without pigmentation are selected as possible haploids (Figure 9.9).

The endosperm tissue is triploid, being composed with a haploid set of chromosome from the male parent and a diploid set from the female parent. The embryo is diploid. As the marker is dominant, in a cross it is always expressed in the endosperm and while in the embryo the marker is absent, since the embryo does not possess a genome from the haploid-inducing line (Belicuas et al., 2007). However, the system is not 100% effective, since the marker may be expressed at a much lower rate in the embryo, giving the impression that the marker is not expressed. In this situation, the breeder may end up selecting it as a false-positive haploid. Its elimitaion could occur later in field, where the plants are more vigorous and better contrasted for pigmentation. This variation in the marker expression is directly associated to the genetic background or in cases in which there are gene inhibitors (C1-I) in the females, a common situation in flint germplasm (Rotarenco et al., 2010). During harvest, it is important that the kernels are dry, otherwise it may be difficult to identify the haploids.

Another phenotypic marker used in maize to identify haploids is the presence of the ligule, controlled by lg_1, lg_2, and lg_3 genes, which confers a more erect-leaf architecture. Plants with erect leaves are diploids and those with more horizontal leaves are haploids (Figure 9.10).

Recently, different markers such as root anthocyanin pigmentation have been used in haploid-inducing lines. Rotarenco et al. (2010) developed new haploid-inducing lines (PHI) from the crosses between MHI, used as a source of desirable traits and high rate of haploid induction, and the Stock 6 line, a source of the B1 and Pl1 marker genes, which do not depend on solar radiation

**With ligule
(hybrid)**

**Without ligule
(haploid plant)**

FIGURE 9.10 Ligule morphological marker in maize for identification of the ploid level. (*Source: Belicuas et al., 2007*)

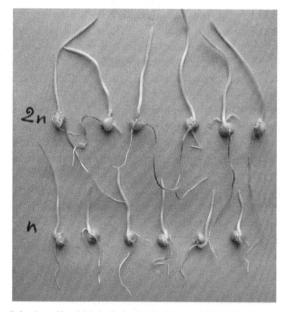

FIGURE 9.11 Selection of haploids by lack of color in roots of seedlings. (*Source: Adapted from Rotarenco et al., 2010*)

to express the anthocyanin pigmentation. In this case, the lack of pigmentation in the roots allows identification of haploid individuals (Figure 9.11).

Chromosome Counting

The most accurate way to confirm the haploid level of an individual is through chromosome counting. This technique does not require costly equipment or chemicals; however, it needs a skilled cytologist to prepare the microscope slides, and is a time-consuming method (Figure 9.12).

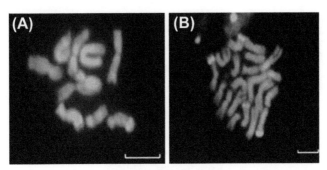

FIGURE 9.12 Chromosome counting. (A) Haploid individual and (B) double-haploid individual. (*Source: Belicuas et al., 2007*)

For chromosome counting, cytological preparations are needed at metaphase. Therefore, the establishment of a routine of meristem root production with high level of mitoses is needed. Germinated seeds are a good source of tissue for those analyses. Other methods for the induction of roots, such as stock rooting in pots or tissue culture, should be considered. Several pre treatments for metaphase accumulation have been described for plants, such as the combination of mitotic-fuse and protein synthesis inhibitors, as well as the use of hydroxyurea to synchronize the meristematic cell division (Cuco et al., 2003).

Colchicine is a mitotic inhibitor commonly used in chromosome duplication and also as a pre treatment for microscope slide preparation. Colchicine bounds to the tubulin and inhibits the formation of microtubules that are part of the fuse fibers, avoiding the migration of the chromosomes to the cell poles (Islam, 2010). This makes it easier to count the chromosomes. For accurate counting it is necessary to discriminate the chromosome from different cells. The Feulgen technique described by Belicuas et al. (2007) is very useful in this regard.

Flux Citometry

Different methods for haploid identification have been developed as the flux citometry. This method is accurate and allows the evaluation of many samples per day; however, it is costly.

The flux citometry is based on the cell division process typical of eukaryotes, composed of the three phases: G1, S, and G2. During the G1 phase, the cell growth period, a diploid cell possesses DNA content that equals 2C, that is, two of each chromosome. In the S phase, there is DNA duplication and in the G2 phase, a new cell growth phase takes place and the nuclear DNA content is 4C. From this phase on, cells are ready for division in two daughter cells (mitosis), when again the nucleus posseses the regular DNA amount, 2C.

The quantification of the ploid level by citometry flux is done by the intensity of the fluorescence emitted by the nucleus stained with DNA-specific fluorochrome. The dominant picks generated in the histograms are proportional to the

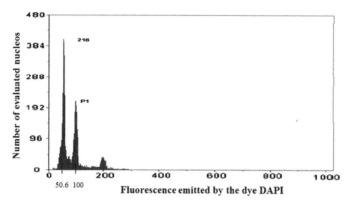

FIGURE 9.13 Histogram from flux citometry of a diploid individual (P1) and from a haploid individual (216). Note the G1 pick of the diploid individual in channel 100 and the G1 pick of the haploid individual in channel 50.6. (*Source: Adapted from Belicuas et al., 2007*)

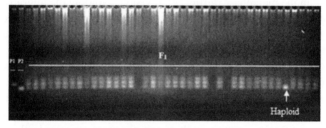

FIGURE 9.14 Identification of androgenetic haploids from crosses between the W23 haploid-inducing line ♀ (P1) and the single cross hybrid BRS1010 ♂ (P2), using the microsatellite primer mmc00ual possesses only a coincident band with a band from the P2 hybrid. (*Source: Adapted from Belicuas et al., 2007*)

DNA amount in the nucleus in the G1 phase (Figure 9.13). The ploid level is determined by comparing the G1 pick in the histogram to a control (Dolezel, 1997).

Molecular Markers

Molecular markers, especially the microsatellites, are useful tools in identifying haploids, since they are multi-allelic and codominant. This is an easier, fast, and accurate technique, since it evaluates the genotype of each individual. The inheritance and stability of the microsatellite locos and its codominance allow the identification of heterozygous individuals (Figure 9.14).

DOUBLE-HAPLOID PRODUCTION

After the identification of the haploid embryos or plants it is necessary to duplicate their chromosome number. All protocols for all crop species are based on colchicine. However, as colchicine is a strong carcinogenic substance, many

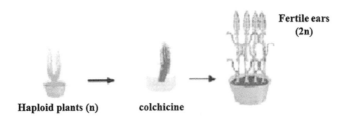

**Fertile ears
(2n)**

Haploid plants (n)   **colchicine**

FIGURE 9.15   Simplified scheme for chromosome duplication and production of DH in wheat.

laboratories are developing less toxic alternatives. Some substances in the dinitroaniline group, such as trifluraline (Kato, 1997), orizaline (Binsfeld et al., 2000), pendimetaline (Zhou et al., 2009), and the organophosphate group, such as amiprophos-metil (Binsfeld et al., 2000), among others, are suggested by Dhooghe et al. (2011) as promising for chromosome doubling.

As previously described, colchicine inhibits the polymerization of the mitotic fuse in the metaphase during cell division, avoiding the migration of the chromosomes to the cell poles, resulting in chromosome duplication after cell division. After treatment with colchicine, the plants are acclimated and finally have their seeds harvested (Figure 9.15).

In the case of haploid maize duplication, seeds are germinated in a moistened germination paper (Figure 9.16A). After the radicule emission (Figure 9.16B), a small cut is done on the seedling coleoptile for better colchicine infiltration (Figure 9.16C). For cochicine treatment, the roots are dried in tissue paper and then placed on a Becker or a tray, making sure the roots are completely under the colchicine solution. In the duplication solution in method I (Deimling, 1997), colchicine is used at 0.06% concentration along with 0.5 to 0.75% of DMSO (dimethyl sulfoxide), leaving the seedlings in the solution for 12 to 17 hours. After this treatment the roots are dried and transplanted to trays with a substract of 1:1 of soil and vermiculite (Figure 9.16D). Water mist should be applied on the plants frequently to avoid dehydration.

After this period, the plants are taken to the field (Figure 9.17A), where seed is set for the double-haploid lines. In maize, about 30% of the haploid plants treated with colchicine may be self pollinated to set seeds (Figure 9.17B), resulting in 100% homozygous plants. Those individuals are known as first generation double haloids. As the seed amount per plant is small they are self-pollinated to produce the second generation double haploids. In this stage the plants have little vigor and are more susceptible to diseases, allowing breeders to select individuals with more desirable traits, leaving only more promising lines for the test crosses.

In the case of wheat, after embryo rescue and transfer to the P2 medium the haploid plants are developed by the maize chromosome elimination, during the first cell division. When the seedlings are 1 to 2 cm long they are transferred to a growth chamber up to the two- to three-leaf stage. Then, those plants are transferred to a tray with a substrate of 1:1 soil and vermiculite up to the four- to six-leaf stage, according to Jobet et al. (2003). At this stage roots are washed

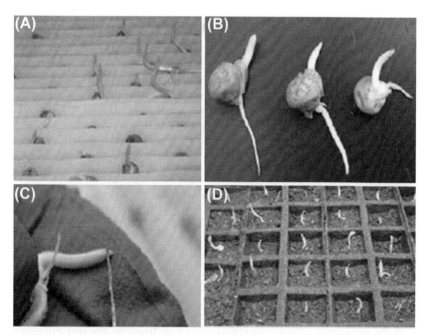

FIGURE 9.16   Chromosome duplication. (A) Germination of haploid seedlings, (B) ideal size for coleoptile cut, (C) coleoptile cut, and (D) transplant to growing trays.

FIGURE 9.17   Maize double haploid in a Monsanto field. (A) Planting and (B) selfing pollination of plants. (*Source: Monsanto*)

and placed in a colchicine solution of 500 mg/L of 2% DMSO, for 6 h at 22 °C under constant aeration. At that time the plants are transferred to pots with fertilized soil and handled until harvest to obtain the double-haploid seeds.

DOUBLE HAPLOIDS IN BREEDING

The value of DH in plant breeding is in the rapid production of homozygous lines. Those homozygous lines are evaluated as possible cultivars in the case of self-pollinated crops or as hybrid parents in the case of cross-pollinated crops.

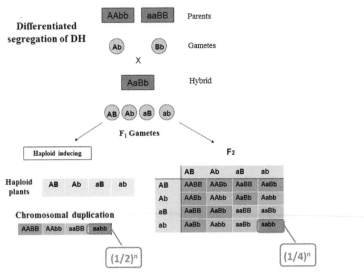

FIGURE 9.18 Differentiated segregation of DH in F_2 and in an ordinary cross.

In the traditional process to obtain homozygous lines, initially the cross among the parents is done to obtain the F_1 population. The F_1 plants are selfed to obtain the F_2 segregating generation. From this point on several different breeding methods are used according to the objectives of the program and the breeder's preference (Borém and Miranda, 2013).

In F_2, the selection is easier for major dominant genes than recessive genes. However, homozygote genotypes with recessive alleles are at the proportion of $(1/4)^n$ of the population, where n is the number of segregating genes. With the chromosome duplication of a F_1 haploid individual, this proportion is $(1/2)^n$ and, therefore, the possibility to obtain a homozygous genotype with recessive alleles is much larger (Figure 9.18). This skewed segregation has other beneficial aspects, such as a better exploitation of the genetic variability and increases of selection efficiency, once homozygous plants possess maximum genetic additive variance and the dominance effects and epistasis are cancelled.

It is important to recognize that the number of double-haploid individuals needed to represent a certain cross is smaller than that used when using the pedigree or the bulk methods, since the F_1 gametes are sampled instead of gametes from F_2 individuals. The number of F_1-plant gametes necessary to constitute a sample can be estimated by the square root of the population size that would ordinarily be required.

Although double-haploid individuals may be directly submitted to evaluation, it is better to multiply their seeds, by self-pollination, since the first generation double-haploid plants may show some epigenetic variations due to the adverse effects of the chemicals used in the chromosome duplication or in the tissue culture phase.

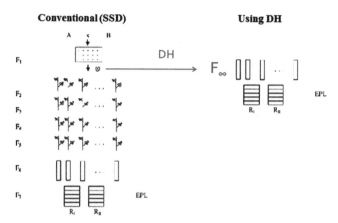

FIGURE 9.19 Inbred production through SSD and double haploids. (*Source: Adapted from Borém and Miranda, 2013*)

Self-Pollinated Crops

For the autogamous species the objective is to obtain homozygous lines, which after agronomic evaluation are released as cultivars. For soybean, the main method of inbred line development is SSD (Borém and Miranda, 2013). In this method, it is necessary for four to six generations of self-pollinations to obtain the homozygous lines. When using DH technology, in just a single generation homozygous lines can be obtained (Figure 9.19). The time saved with the use of DH with annual species is not as rewarding as in perennial species.

Cross-Pollinated Species

In cross-pollinated species the DH can be used in two ways: (1) in the production of inbred lines to be used as hybrid parents or even in recycling lines and (2) to improve the genetic gain in recurrent selection programs.

The first use is similar to what was described in self-pollinated crops. This technique is widely used in the maize breeding industry, especially using the haploid-inducing lines. The latter has been studied by many groups around the world, but with very few reports in practical programs. With DH, available breeding methods, such as the recurrent selection, could be more attractive to the breeding industry, due to the maximization of genetic gains with the use of completely homozygous parents in the recombination phases. Additionally, other methods such as reciprocal recurrent genomic selection could also be used (Figure 9.20), through genome-wide selection (Fritsche-Neto et al., 2012).

In this method the DH could be used in two phases. The first phase would be the production of the improved lines. The second phase would be after the cycles of the genome-wide selection within the population, where it would be required to obtain the inbred lines to estimate and identify the markers that maximize the accuracy in the hybrid selection (Fritsche-Neto et al., 2012).

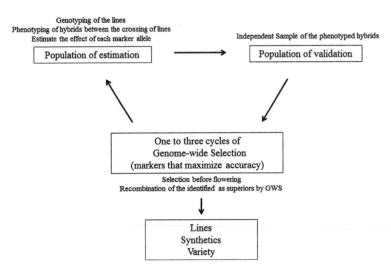

FIGURE 9.20   Recurrent genomic selection. (*Source: Adapted from Fritsche-Neto et al., 2012*)

ADVANTAGES AND DISADVANTAGES IN THE USE OF DOUBLE HAPLOIDS

For a generalized adoption of DH in breeding programs the following would be required: (1) a method for production of a large number of DH in an economical and technically compatible fashion with the available resources; (2) that the method works finely with all germplasms (absence of genotype × method interaction); (3) that the obtained DH be a sample of all variability from the individuals from which they were produced (no gametic selection); and (4) the existence of an efficient chromosome duplication method (Borém and Miranda, 2013).

Table 9.3 presents the main advantages and disadvantages of DH in plant breeding.

APPLICATION ON THE GENOMIC STATISTICS

The power to detect (or map) a QTL (*quantitative trait loci*) depends on several factors, among them the segregating population type and size (Lynch and Walsh, 1998). In this regard, the use of DH populations may result in significant gains in time, especially for perennial species or those with a long generation time or with any difficulty in generating the homozygous lines, as in the case of citrus, passion fruit, and coffee (Figure 9.21).

Recombinant inbred lines (RILs) produced through double haploids, for use in QTL mapping, have the advantage of the reduction of undesirable recombinations between the target region and the marker, as it happens in each generation of sexual reproduction. Other advantages are the rapid production of

TABLE 9.3 Main Advantages and Disadvantages of DH in Plant Breeding

Advantages	Disadvantages
Rapid production of homozygous lines	Low efficiency
Immediate identification of mutants	Segregation can be biased
Differentiated segregation: a smaller number of individuals are needed to produce the desired recombinants	Low recombination
Purification and recycling of inbred lines	Risk of haploid-inducing gene escapes
Elimination of lethal and deleterious genes	Germplasm is selected for *in vitro* regeneration
Improve variability and efficiency in selection	Variability without previous selection (little chance to discard inferior individuals)
Shorten the time to release a new cultivar	The DH may not represent an actual gain in shortening the breeding program

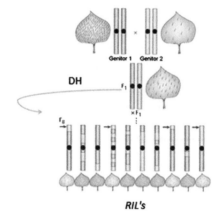

FIGURE 9.21 Production of recombinant inbred lines (RILs) through double haploids, for QTL mapping.

the RILs and the smaller possibility of unintentional selection on the mapping population, since the RILs are not exposed to the several generations of selfing as with the traditional RILs.

To obtain this type of population it is necessary to use protocols for double-haploid production, which are currently available only for a small number of crop species. Another disadvantage is the possibility of somaclonal variation due to the regeneration and the chromosome duplication processes, resulting in a skewed gametic sample of the parents, which is undesirable in QTL mapping.

ECONOMICAL ASPECTS

Considering all aspects addressed in the chapter, it is noteworthy to point out that plant breeding should take advantage of the use of DH. Gomez-Pando et al. (2009) have shown that barley lines derived through DH have high grain yield potential and excellent nutritional quality. Additionally this technology has been proven competitive on cost analyses in relation to the conventional methodologies. Evidence of this is that most of the private breeding programs are adopting this technique, especially because it saves time in the development of new cultivars.

Finally, although the DH technique can be used for many crops it does not work for some important species.

REFERENCES

Belicuas, P.R., Guimarães, C.T., Paiva, L.V., Duarte, J.M., Maluf, W.R., Paiva, E., 2007. Androgenetic haploids and SSR markers as tools for the development of tropical maize hybrids. Euphytica 156, 95–102.

Bidmeshkipour, A., Thengane, R.J., Bahagvat, M.D., Ghaffari, S.M., Rao, V.S., 2007. Production of haploid wheat via maize pollination. Journal of Sciences 18(1), 5–11.

Binsfeld, P.C., Peters, J.A., Schnabl, H., 2000. Efeito de herbicidas sobre a polimerização dos microtúbulos e indução de micronúcleos em protoplastos de *Helianthus maximiliani*. Revista Brasileira de Fisiologia Vegetal 12(3), 263–272.

Bjornstad, A., 1989. Protocol for barley anther culture. [S.l.]: Department of Genetics and Plant Breeding. Agricultural University of Norway, p. 74.

Blakelsee, A.F., Belling, J., Farhnam, M.E., Bergner, A.D., 1922. A haploid mutant in the Jimson weed, Datura stramonium. Science 55, 646–647.

Borém, A., Miranda, G.V., 2013. Melhoramento de plantas, sixth ed. Editora UFV, Viçosa. p. 523.

Bylich, V.G., Chalyk, S.T., 1996. Existence of pollen grains with a pair of morphologically different sperm nuclei as a possible cause of the haploid-inducing capacity in ZMS line. Maize Genetics Cooperation Newsletter 70, 33.

Chalyk, S.T., 1994. Properties of maternal haploid maize plants and potential application to maize breeding. Euphytica 79, 13–18.

Chalyk, S.T., 1999. Creating new haploid-inducing lines of maize. Maize Genetics Cooperation Newsletter 73, 53–54.

Chalyk, S., Baumann, A., Daniel, G., Eder, J., 2003. Aneuploidy as a possible cause of haploid-induction in maize. Maize Genetics Cooperation Newsletter 77, 29.

Chase, S.S., 1974. Utilization of haploids in plant breeding: breeding diploid species. In: Kasha, K.J. (Ed.), Haploids in Higher Plants: Advances and Potential, pp. 211–230. Proceedings of the First International Symposium, Guelph, Canada.

Chen, J.F., Cui, L., Malik, A.A., Mbira, K.G., 2011. In vitro haploid and dihaploid production via unfertilized ovule culture. Plant Cell and Tissue Culture 104, 311–319.

Chuang, C.C., Ouyang, T.W., Chia, H., Chou, S.M., Ching, C.K., 1978. A set of potato media for wheat anther culture. In: Proceedings of the Symposium on Plant Tissue Culture. Science Press, Peking, pp. 51–56.

Clauszn, R.R., Mann, M.C., 1924. Inheritance in Nicotiana Tabacum: V. The occurrence of haploid plants in interspecific progenies. Proceedings of the National Academy of Sciences 10 (4), 121–124.

Coe, E.H., 1959. A line of maize with high haploid frequency. The American Naturalist, Chicago 93, 381–382.

Coe, E.H., Neuffer, M.G., 2005. Darker orange endosperm color associated with haploid embryos: Y1 dosage and the mechanism of haploid induction. Maize Genetics Cooperation Newsletter 79, 7.

Cuco, S.M., Mondin, M., Vieira, M.L.C., Aguiar-Perecin, M.L.R., 2003. Técnicas para a obtenção de preparações citológicas com altas frequências metáfases mitóticas em plantas: Passiflora (Passifloraceae) e Crotalaria eguminosae). Acta Botânica Brasilica 17 (3), 363–370.

Deimling, S., Röber, F., Geiger, H.H., 1997. Methodik und genetik der in-vivo-haploideninduktion bei mais. Vortr Pflanzenzüchtung 38, 203–204.

Dhooghe, E., Van Laere, K., Eeckhaut, T., Leus, L., Huylenbroeck Van, J., 2011. Mitotic chromosome doubling of plant tissues in vitro. Plant Cell Tissue Organ Culture 104, 359–737.

Dolezel, J., 1997. Applications of flow cytometry for the study of plant genomes. Journal of Applied Genetics 38 (3), 285–302.

Dunwell, J.M., 2010. Haploids in flowering plants: origins and exploitation. Plant Biotechnology Journal 8 (4), 377–424.

Fritsche-Neto, R., Resende, M.D.V., Miranda, G.V., Do Vale, J.C., 2012. Seleção genômica ampla e novos métodos de melhoramento do milho. Ceres 59, 794–802.

Gaines, E.F., Aase, H.C., 1926. A haploid wheat plant. American Journal of Botany 13 (6), 373–385.

Gernand, D., Rutten, T., Varshney, A., Rubtsova, M., Prodanovic, S., Bruß, C., et al., 2005. Uniparental chromosome elimination at mitosis and interphase in wheat and pearl millet crosses involves micronucleus formation, progressive heterochromatinization, and DNA fragmentation. Plant Cell 17, 2431–2438.

Gomez-Pando, L.R., Jimenez-Davalos, J., Eguiluz-De la Barra, A., Aguilar-Castellanos, E., Falconí-Palomino, J., Ibañez-Tremolada, M., 2009. Estimated economic benefit of double-haploid technique for Peruvian barley growers and breeders. Cereal Research Communications 37 (2), 287–293.

Guha, S., Maheshwari, S.C., 1964. In vitro production of embryos from anthers of Datura. Nature 204, 497.

Guha, S., Maheswari, S.C., 1966. Cell division and differentiation of embryos in the pollen grains of Datura in vitro. Nature 212, 202–297.

Henshaw, G.G., O'Hara, J.F., Webb, K.J., 1982. Morphogenetic studies in plant tissue culture. In: Yeoman, M.N., Truman, D.E.S. (Eds.), Differentiation in vitro. Cambridge University Press, Cambridge, pp. 231–251.

Hu, H., 1986. Variability and gamete expression in pollen-derived plants in wheat. In: Han, H., Hongyuang, Y. (Eds.), Haploids of higher plants in vitro. Springer Verlag, Berlin, pp. 67–68.

Hu, H., 1997. In vitro induced haploids in wheat. In: Velleiux, J.S. (Ed.), In vitro Haploid Production in Higher Plants. Kluwer Academic Publishers, Dordrecht, Boston, pp. 73–97.

Hu, S., 1990. Male germ unit and sperm heteromorphism: the current status. Acta Botanica Sinica 32, 230–240.

Huang, B., Sheridan, W.F., 1996. Embryo sac development in the maize indeterminate gametophyte 1 mutant: abnormal nuclear behavior and defective microtubule organization. The Plant Cell 8, 1391–1407.

Islam, S.M.S., 2010. The effect of colchicine pretreatment on isolated microspore culture of wheat (Triticum aestivum L.). Australian Journal of Crop Science 4, 660–665.

Jobet, J., Zúñiga, J., Quiroz, H.C., 2003. Plantas doble haplóides generadas por cruza de trigo × maíz. Agricultura Técnica 63 (3), 323–328.

Kato, A., 1997. Induced single fertilization in maize. Sex Plant Reproduction 10, 96–100.

Kermicle, J.L., 1969. Androgenesis conditioned by a mutation in maize. Science 166, 1422–1424.

Lashermes, P., Beckert, M., 1988. Genetic control of maternal haploidy in maize (*Zea mays* L.) and selection of haploid inducing lines. Theoretical Applied Genetics 76, 405–410.

Lashermes, P., Gaihard, A., Beckert, M., 1988. Gymnogenetic haploid plants analysis for agronomic and enzymatic markers in maize (*Zea mays* L.). Theoretical and Applied Genetics 76, 570–572.

Lynch, M., Walsh, B., 1998. Genetics and analysis of quantitative traits. Sinauer Sunderland, Massachusetts. p. 980.

Ma, H., Busch, R., Riera-Lizarazu, O., Rines, H., 1999. Agronomic performance of lines derived from anther culture, maize pollination and single seed descent in a spring wheat cross. Theoretical Applied Genetics 99, 432–436.

Moura, M.M., Rosa, L.M.P., Oliveira, L.A., Figueira, D.P., Martins, P.K., Franco, F.A., et al., 2008. Otimização de plantas duplo-haplóides de trigo através da hibridação trigo × milho: seleção de híbridos polinizadores e meios de cultura. Resumos do 54° Congresso Brasileiro de Genética, Salvador-BA, p. 221.

Murovec, J., Bohanec, B., 2012. Haploids and doubled haploids in plant breeding. In: Abdurakhmonov, I. (Ed.), Plant Breeding. InTech. Available from: <http://www.intechopen.com/books/plant-breeding/haploids-and-doubled-haploids-in-plant-breeding>.

Nanda, D.K., Chase, S.S., 1966. An embryo marker for detecting monoploids of maize (*Zea mays* L.). Crop Science 6, 213–215.

Röber, F.K., Gordillo, G.A., Geiger, H.H., 2005. *In vivo* haploid induction in maize performance of new inducers and significance of doubled haploid lines in hybrid breeding. Maydica 50, 275–283.

Rotarenco, V., Dicu, G., State, D., Fuia, S., 2010. New inducers of maternal haploids in maize. Maize Genetics Cooperation Newsletter 84.

Tyrnov, V.S., Zavalishina, A.N., 1984. Inducing high frequency of matroclinal haploids in maize. Doklady Akademii Nauk SSSR 276, 735–738.

Veilleux, R.E., 1994. Development of new cultivars via anther culture. Hortscience 29 (11), 1238–1241.

Yonezawa, K.O., Nomura, T., Sasaki, Y., 1987. Conditions favoring doubled haploid breeding over conventional breeding of self-fertilizing crops. Euphytica 36, 441–453.

Zhang, Z., Qiu, F., Liu, Y., Ma, K., Li, Z., Xu, S., 2008. Chromosome elimination and *in vivo* haploid production induced by Stock 6-derived inducer line in maize (*Zea mays* L.). Plant Cell Reproduction 27, 1851–1860.

Zhou, X.J., Cheng, Z.H., Meng, H.W., 2009. Effects of pendimethalin on garlic chromosome doubling *in vitro*. Acta Botanica Boreali-Occidentalia Sinica 29 (12), 2571–2575.

Chapter 10

Tools for the Future Breeder

Valdir Diola,[a] Aluízio Borém,[b] and Natália Arruda Sanglard[a]

[a]Federal Rural University of Rio de Janeiro, Seropédica-RJ, Brazil, [b]Federal University of Viçosa, Viçosa-MG, Brazil

INTRODUCTION

Currently, the genetic improvement of plants demands the selection process to be more efficient and to obtain better results in a short period of time. However, the productivity of most of the cultivations has almost reached a plateau. Many cultivations show high genomic complexity; polyploidies, aneuploidies, recombinations, structural chromosome alterations (translocation, inversion, deletion, and insertion), multiple alleles, interaction of QTLs and eQTLs, and high gene variation make it difficult to obtain cultivars more stable and less vulnerable to environmental alterations. The agronomic characteristics of interest are predominantly of a quantitative nature and are subjected to a pronounced environmental effect, hampering genetic improvement processes. Until now, the current study has prioritized production and quality phenotypic characters, selecting genotypes in only one field level. Given the increase in the difficulty of obtaining better varieties, genetic typing was used, which was very useful when backed up by phenotyping. These strategies increase discriminatory power in the early selection of progenies and reduce attainment period for new varieties. There are many genotyping methodologies, each one having its own application characteristics and specificities, and the breeder must know and become acquainted with the most appropriate choice. Nowadays, genotyping techniques are no longer limited, due to automatization and subcontracting. It is believed that, in the future, phenotyping will be one of the most limiting factors for genetic improvement, especially for large populations.

Biotechnology has also revealed itself to be very useful for basic genetics and in the attainment of transforming varieties. For almost two decades, the area cultivated with these varieties has increased steadily and species such as soy and corn, genetically modified cultivars, have already used up a larger area than conventional ones.

This chapter will take a generalist approach to the contributions that new methodologies have already brought and will be able to bring to the improvement of plants in the very near future.

Biotechnology Applied to Plant Breeding. http://dx.doi.org/10.1016/B978-0-12-418672-9.00010-6

GENOMIC TOOLS

Genomics is understood to be the activities correlated with the study and application of techniques comprising DNA (structural genome) and RNA (functional genome). Despite the fact that the techniques for both genomes show high operational similarity, the technological application and its intended purpose in most of the cases are very distinct. Structural genome techniques analyze the transmission of inheritable characteristics, epigenetic regulation (transient), somatic variation, and cytogenetic alterations (constitutive). Functional genome techniques relate to transcriptional and post-transcriptional regulations comprising RNA: hnRNA (heterogeneous RNA), mRNA (messenger RNA), rRNA (ribosomal RNA), tRNA (transfer RNA), snRNA (small nuclear RNAs), among others. They are intended to study the behavior of expression which may embrace a specific gene, even the complete transcriptome under certain environmental stimulus in time and intensity. Generally, transcription levels correlate with the respective protein translations and with the intensity of the encoded phenotype expression. There are several regulation levels between hereditary nature and phenotypic expression (Figure 10.1), and genetic factors can be temporarily or permanently overregulated.

In most of the cases, there is a need to integrate structural genomic, transcriptomic, proteomic, and/or metabolomic techniques, since most of the gene expression control occurs at promoter level (effect on transcription) and to a lesser extent on the coding region (effect on protein activity).

TECHNIQUES APPLIED TO STRUCTURAL GENOME

Genotyping via DNA is nowadays widely applied to the genetic improvement of plants. Initially, with the discovery of DNA restriction enzymes, molecular markers such as RFLP (*restriction fragment length polymorphism*) could be obtained (Botstein et al., 1980). However, the technique that contributed most to genomics was DNA-molecule amplification. Initially, it was very laborious, since *DNA polymerase III* lost its stability in each polymerization cycle. Mullus and Falona, in 1987, provided a relevant contribution to PCR (*polymerase chain reaction*) automatization, discovering thermotolerant *DNA polymerase III*, allowing consecutive cycling at high temperature without denaturing (~94°C for 2 hours). This fact allowed the performance of PCRs in automatic thermal cyclers.

DNA fragmentation with endonucleases is also applied to genetic cloning and transformation, as well as the recombination and joining of enzymes of *in vitro* DNA molecules (*DNA ligase*): e.g., inserts in plasmids, vectors of the molecules of interest. The linkage and combination of these techniques have improved the processes for obtaining the most diverse enzymes, which became heterologous, in most of the cases from bacteria of other strains. The production of the enzymes used in molecular biology has been optimized, reducing costs

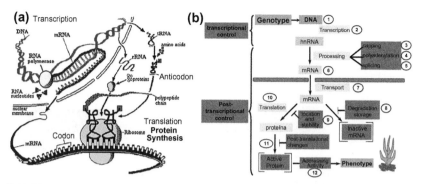

FIGURE 10.1 Detailing from the central dogma of the molecular biology. (a) Process for transcription of DNA into RNA and translation of mRNA into protein. (b) Control of gene expression in eukaryotes: (1) epigenetic control: methylation, acetylation, and phosphorylation of histones and other proteins and DNA, (2) control at the transcriptional level: changes in assembly conditions, processivity and/or stop the RNA polymerase complex, (3) capping: changes in the activity of the complex 7-methyl-guanylyl-transferase on the 5′ end of hnRNA, (4) polyadenylation: activity and polyadenylation complex formation and activation of polyA-polymerase, (5) splicing: timely correct assembly of the spliceosome snRNP and the removal of introns, (6) addressing the mRNA: RNA binding protein to protect and mRNA export, (7) transport: addressing and protection of mRNA to ribosomes in the cytosol, (8) degradation or storage: interaction with small RNA followed by degradation or folding temporary, (9) location and stability: correct target mRNA and suitable environmental conditions for the cytosol, (10) translation: assembly of complex translation (initiation and elongation), (11) post-translational changes: folding, phosphorylation, oxidation, and protein sulfurization, (12) addressing and activity of protein: protein binding addressing, encapsulation, glycosylation, oxidation, and interacting protein/protein, (13) regulating in the substrate level: cofactors, substrates, and products (metabolism), and functional activity of the protein or enzyme.

and allowing fast progress of molecular techniques. Such techniques still play an expressive role in genomics: genotyping, sequencing, and genetic transformation.

The progress achieved in biotechnology and genomic techniques, such as DNA markers, represented a notable contribution as tools for the analysis, selection, and breeding of improved varieties (Borém and Teixeira, 2009). When applied in the early stages of genetic improvement, in progeny selection of first generations and of genitors, it facilitates and reduces time in the identification of genotypes with the desired characteristics.

RFLP molecular markers were first obtained in the 1970s, and the possibility of a DNA genotype-based selection instead of a phenotype-based one was soon identified. Techniques were implemented in order to obtain more specific and discriminatory molecular markers that can be divided as derived from restriction followed or not followed by amplification and derived exclusively from PCR. RFLP (*restriction fragment length polymorphism*) is obtained by restriction. Markers derived from restriction/amplification are AFLP (*amplified fragment length polymorphism*) and CAPS (*cleaved amplified polymorphic sequence*);

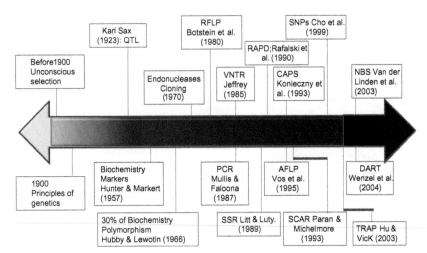

FIGURE 10.2 Main methodologies of genetics; molecular markers and researchers who contributed to the generation of new technologies with application in plant breeding.

while those exclusively obtained by PCR are RAPD (*random amplified polymorphic DNA*), SCAR (*sequence characterized amplified regions*), SSR (*simple sequence repeats*), and SNP (*single nucleotide polymorphism*) (Figure 10.2), the last marker being the most used in plant breeding programs.

All these markers are still being used, to a greater or lesser extent, depending on their purpose. New markers emerge, however more specific, in order to meet the main needs of the breeding programs. When species or a related genome is known, collinearity (synteny) is used, facilitating the generation of primers in order to discriminate the desired phenotypes.

The use of these tools is derived from the system's applicability and operability, as well as from the physical infrastructure of laboratories. However, for efficiency and specificity reasons, the use of a certain technology may take precedence over others. Most recently, SNPs and sequencing-derived markers (Table 10.1) are the most prominent ones (Figure 10.3), since small- and large-scale operationalization and automation have become possible, at affordable costs, to the majority of laboratories.

The attainment of a large number of markers has become possible by using an automated sequencer (with 96 or 384 samples per test), facilitating the application of genotyping techniques on a large scale. Since the advent of large-scale sequencing, around 2004, the first pyrosequencing (Roche-454) events occurred and a new genotyping era began. With these techniques, generation power of molecular markers was amplified many times regarding the use of regular PCRs followed by electrophoresis. Nowadays, more structured breeding programs request some thousands of markers, which does not yet justify the acquisition of

TABLE 10.1 Main Technical Characteristics to Obtain Molecular Markers and Genomic Basic Applicability in Plant Breeding Programs

Techniques	Technical principles	Recommendations from applications	Advantages	Disadvantages
RFLP	Restriction enzyme digestion of DNA followed by electrophoresis and binding with specific probes	– Isolation of regions of interest – Genetic mapping – Genotyping	– High specificity of hybridization probes to loci of interest	– Technical and operational difficulties, requiring ability professional
RAPD	Random amplification of sequences located between pairs of 10 bp primers	– Genotyping – Genetic mapping	– Application ease	– Low reproducibility between laboratories
SSR	Amplification of regions with repetitive sequences of DNA	– MAS – Genotyping – Genetic mapping	– After the identification of polymorphism, available for use	– Need knowledge of the regions that are intended to discriminate
AFLPs	Restriction enzyme genome, connection adapters, and PCR primers with selective bases at two stages of amplification	– Genetic mapping – Genotyping requiring large number of molecular markers	– High capacity of generating loci randomly distributed in the genome	– High cost of implementation – Need for good technology- Generation of a secondary marker for application in MAS
SCARs	Amplification of a polymorphic region derived from a molecular marker (marker secondary)	– Positional cloning – MAS	– High discriminatory power of polymorphic sequences – Reproducibility – Operability	– Dependent on primary marker

(continued)

TABLE 10.1 Continued

Techniques	Technical principles	Recommendations from applications	Advantages	Disadvantages
CAPs	Fragmentation of the sequence of a marker and reamplification	– Specific genotyping – MAS	– Possibility to better discriminate the polymorphic locus	– Dependent on primary marker
SNPs	Amplification of regions containing simple changes based	– Specific genotyping and large scale – MAS	– Exploration of the abundance of polymorphism in specific loci	– requirement of prior knowledge of sequences and SNPs
INDELs	Amplification of regions of interest containing numerical variation in the base sequence	– Specific genotyping – MAS	– Ease of detection of polymorphism	– Low abundance of indels in the loci of interest – Knowledge of the sequences
NBS-profiling	Amplification of a fragment derived from restriction using an adapter primer and NBS	– Linked polymorphism genotyping for disease resistance	– Application in the study specific resistance	– Previous knowledge of the sequences of the NBS-LRR genes for obtaining the fixed primers
TRAP	Amplification of specific regions using a fixed primer annealing combined with a random	– Study of the polymorphism near the region of interest	– Identification of genetic variability co-segregating with traits of interest	– Previous knowledge of sequences and their relationship with the phenotype – Requires a secondary marker for use in MAS
DART	Marker that is based on hybridization of probes (primers) with sequences of interest detected by fluorescence levels	– Large-scale genotyping – MAS	– Obtaining large numbers of markers or discrimination of large numbers of genotypes in short time	– High demand: technology, operational technology, initial financial resources, and professional ability

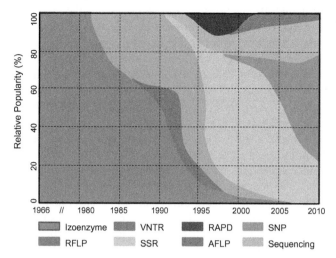

FIGURE 10.3 Popularity of the use of molecular markers (including biochemical marker iso-enzyme) during the period from 1966 to 2010, according to publications in international journals.

this sort of equipment, making the outsourcing of services a very common alternative, especially due to the quantities undertaken. This methodology represents a significant impact on the reduction of human interference and on efficiency maximization. No matter how advanced this technology might be its base is grounded on the principles of PCR combined with techniques of basic sequencing and, sometimes, electrophoresis.

TECHNIQUES APPLIED TO FUNCTIONAL GENOME

Up to mid-1990, techniques involving functional genome did not show relevant expression; however, they are the focal point of most studies nowadays. They explore alterations of level on gene expression composing functional genome; i.e., they involve the study of transcription level difference over time under certain stimulus. The main advantage of functional genome study is the analysis of a reduced genome; i.e., only genome expressed regions are studied (Figure 10.4; from ESTs to mRNA full length). Expressed genome represents less than 1.5% of DNA's relative size. Phenotypic characteristics depend on this expressed fraction facilitating a genotyping approach related to phenotype. Transcription is the result of complex *RNA polymerase* activation; it depends on the environment stimulus and is often very specific to each genotype and even to each individual. Some characteristics remain more stable and are inheritable, generally related to structural genome alterations occurring in the promoter region. They affect the transcription level because alterations on promoter sequences may affect *RNA polymerase* complex assembly. Usually, there is a reduction of protein final quantity, but the coding sequence remains preserved and protein activity is unaltered. If mutations affect coding regions, the transcription level is generally preserved, but protein codons are altered in order of basic amino acid constitution and they probably encode into a protein with partial activity or produce a deleterious effect (functionless). These mutations cause phenotype manifestation as long as its character is dominant or in recessive homozygosis. Molecular techniques are still not efficient in the detection of these random mutations, but type-TRAP (*target region amplification polymorphism*) molecular markers are specified the most for identification of the same adjacent regions to the gene of interest, since it allows the identification of such mutations by differential amplification among different genotypes.

TRAP and epigenetic techniques (which use DNA bisulfitation) do not depend on the attainment of RNA molecules. In both cases, cDNA (complementary DNA) attainment (Figure 10.4) can be the basis of the research or complementary to genome studies. There are many alternatives to the selection of functional genome techniques that may be applied to genetic improvement. Some of them are listed in Table 10.2. It is worth highlighting that most of these techniques involve a cDNA synthesis in order to keep molecule stability, since

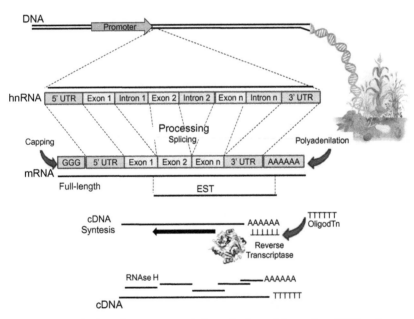

FIGURE 10.4 The dependent gene transcription activation and interaction of RNA polymerase complex on the promoter for the synthesis of heterogeneous RNA (hnRNA) followed by processing involving the removal of introns, capping, and polyadenylation. A full-length transcript or mRNA product is the complete complementary synthesis from the cap (7-methylguanosine), different from which cDNA synthesis from polyA is, which basically applies in obtaining ESTs which include the partial mRNA.

TABLE 10.2 Techniques for the Study of Functional Genomics and Applications in Plant Breeding

Techniques	Technical Principles	Recommendations from Applications	Advantages	Disadvantages
Northern blot	mRNA isolation, electrophoresis, transference to membranes and hybridization with specific probes	– Semi-quantitative analysis of transcript levels of specific genes in treatment	– No demand for high technologies	– Low productivity – Demand for specific hybridization probes
ESTs library	Isolation of mRNA and synthesis of double-strand cDNA (though not necessarily) and cloning	– Studies of expressed sequences in specific treatment	– Maintenance of a stock of clones which can be appealed when necessary	– High costs – Dependent sequencing – Very laborious
Full length library	Similar to ESTs, but the complete genes and cDNA synthesis is performed from the cap	– Studies that require knowledge of the complete gene, especially for use in genetic transformation	– Allows complete knowledge of the gene sequence	– Need for highly qualified professional to obtain the full length – mRNA excellent quality
PCR subtractive suppressive hybridization	Obtaining mRNA genotype contrasting or different treatment, digestion of cDNA subtraction hybridization and PCR	– Identification and selection of differentially expressed genes	– Only a fraction differential is evaluated	– Operationalizing quite complex and laborious
qRT-PCR (PCR real time)	Amplification of a number of genes of interest to detect and quantify the intensity level of transcription	– Quantitative analysis of gene expression	– Ease of execution and analysis – High performance; equipment is often improved	– Operation difficult and cost relatively high – Demand of homogeneity in the samples and PCR reactions

(continued)

TABLE 10.2 Continued

Techniques	Technical Principles	Recommendations from Applications	Advantages	Disadvantages
Functional markers	Same principles applied to structural genomics	– Analysis of polymorphisms linked to phenotypic traits of interest	– Reducing the sample size compared to the whole genome (approximately 1–2%)	– Instability is affected by the environment and reproducibility is variable according to the intensity of treatment
RNASeq	Obtaining the mRNA and next-generation sequencing	– Transcriptome analysis under specific treatment	– Quantitative determination of gene expression and knowledge of the genetic variability of genes	– Requirement for extensive knowledge of bioinformatics tools
Macro- and micro-arrays	Analysis of a large set of genes based on hybridization and fluorescence detection	– Genotyping large populations – Quantitative analysis of the expression of a set of genes	– Operation and automated techniques for gene expression and genotyping	– Relatively high cost because it depends on chip hybridization – Dependent on genetic material of high quality

mRNA naturally has a short half-life, being rapidly degraded if storage cap conditions are not adequate.

In most of the cases, functional genome techniques depend on other techniques for confirmation, especially on those exploring structural genome. Thus, they can be applied to plant breeding in the form of genetic markers, since they are very close to or among candidate genes or target genes and they have the likelihood of establishing correlations between phenotype and genotype as their main advantage.

Among the main functional markers are: cDNA-AFLP, SNP, TRAP, NBS, S-SAP, REMAP… (dependent on PCR), and DArT (dependent on hybridization). Among markers dependent on PCR, cDNA-AFLP was widely popularized and represented a significant contribution to plant breeding. However, due to its dependence on a secondary marker and on several techniques used to do so, other techniques have been preferred. Hu and Vick (2003) developed a TRAP technique based on PCR and ESTs sequences that have shown great potential for the detailing of polymorphism close to the target gene region. An NBS profiling marker is widely used in the identification of polymorphism of genes resistant to plant diseases, since they explore the NBS gene variable and conserved domains. However, SNPs are the most accepted markers due to their abundance and high specificity.

The DArT (*diversity arrays technology*) marker shows great potential for the study of genetic and mapping diversity, with a high capacity of polymorphic discrimination and very accurate analyses. DArT technology has advantages over other technologies due to its cost and fast polymorphism detection. It can be automated with the use of microarray machines, mass spectrometer (MALDI-TOF), and a new generation sequencing equipment. NGSs are able to genotype hundreds and thousands of genotypes in a few days with thousands of probes. Among the most used probe technologies are Affymetrix® chips, which are based on short sequences of oligonucleotides. Most recently, biochips (based on protein/DNA interaction) have been widely used. It is a high-performance method which detects interactions and DNA/protein interaction activities, predicts protein function, and its main advantage is centered on the fact that a large number of proteins can be analyzed at the same time.

AVERAGE- AND LARGE-SCALE ATTAINMENT OF MARKERS

The performance of average-scale molecular biology techniques has always been complex, since electrophoretic systems are laborious, especially in genotyping populations with high numbers of individuals. Since the mid-1990s, techniques based on automatic sequencing have been more extensively used, because they allow the genotyping of relatively large populations, usually with more than 1000 individuals. They are based on fast amplification and fragment separation in capillary systems. Some sequencers work with plates of 384 wells or more, allowing multiplex reactions (several primer sets in the same reaction) and thus speed up the procedure. These systems have a wide advantage over traditional electrophoresis (small-scale markers), since separation pattern does not depend on chemical coloration and visualization, reducing operator errors.

Other platforms are also used and technically recommended for SNPs and SSRs, since they have specific characteristics, as follows:

- Pyrosequencing – few SNPs, few samples, lower middle performance and intense work (in more intense works, Barcode is used: multiplex primer marking with *specific* fluorofluors for each genotype).
- Real-Time – TaqMam System – few SNPs, few samples, middle performance (the system is laborious and requests high-class reagents and a large number of primers related to polymorphism).
- Fluidigm, SNPlex, Sequenom – moderate number of SNPs and samples, high performance, with thousands of reactions per day, systems using MALDI-TOF being the most common ones.

- Illumina GA Analyzer – many SNPs, many samples, ultra-high performance (*ultra-high-throughput*) with millions of SNPs per day.
- Affymetrix – many SNPs, many samples, *ultra-high-throughput* (millions of samples per day) (system based on DNA arrangement and hybridization in *chip arrays*).
- Other new generation sequencing platforms – Illumina HiSeq 2500, Solid and others such as nanopore, Helicos, and PacBio (Pacific Biosystem); show parallelism and *ultra-ultra-high-throughput* with the capacity of one genome in a couple of hours.

Small-scale genotyping systems based on real-time PCR (qRT-PCR) have gained a prominent position and popularity due to the costs of operation and equipment, especially in medium and large laboratories. Initially, the TaqMan system was applied to reactions using three primers, two for the amplification and one as a probe, the function of which is to emit fluorescence and to allow multiplex reactions.

In medium- and large-scale genotyping carried out in microarray equipment, mass spectrometers (MALDI-TOF), and recently in NFS (*next generation sequencing*), DNA arrays stand out. Platforms using mass spectrometry depend on plates of 384 wells (to a lesser extent) or on chips, in large scale, with more than 10,000 samples. There are four sample preparation principles (Figure 10.5a): (1) based only on probe hybridization; (2) based on primers and extension hybridization, dependent on PCR; (3) based on hybridization by oligo; and (4) based on cleavage with endonucleases. The analysis is performed according to fragment density (Figure 10.5b). This technology has been used in laboratories that have equipment generally used for proteomic data analysis, which is very efficient in genotyping. However, this technology is limited by the construction of the chips containing the group of genes to be studied. For some species, such as rice, wheat, and corn, these chips are already available, circumstantially reducing the costs.

NGS equipment uses hybridization followed by PCR and produces thousands or millions of markers per day, being considered a large-scale genotyping technology. Some equipment is based on primer hybridization and allele-specific extension, of a lower scale, but more precise, and is massively adopted by researchers.

For technical and financial reasons, large research groups demanding elevated operational capacity motivated some companies to offer sequencing and genotyping services. Nowadays, many companies offer such service and outsourcing, which has become highly advantageous for financial reasons regarding equipment, physical space, profitability, and workability. Many of these companies offer options regarding costs, performance, production quality, and technical platforms.

Genotyping by sequencing (GBS) has been widely applied to genetic improvement of plants in order to obtain more expressive markers: SNPs, SSRs, INDELs, variations on the number of copies, RAPD, AFLP. Other analysis can be performed in NGS: gene expression, DNA methylation, loss

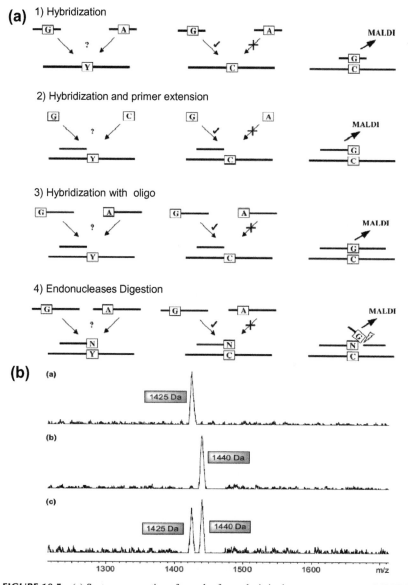

FIGURE 10.5 (a) System preparation of samples for analysis in the mass spectrometer MALDI-TOF. (b) System analysis of mass spectra: (a) homozygous genotype containing the loco with 1425 Da, (b) homozygous genotype containing the loco with 1440 Da, and (c) heterozygous genotype containing the loco with 1425 and 1440 Da.

of heterozygosity, siRNA, RNAi and microRNA, BAC fingerprint, SAGE, allele quantification, sequencing, re sequencing, comparative genomic, *de novo* sequencing, and large fragment sequencing.

The main financial advantages offered by these platforms are:

- Miniaturization: more compact platforms, reduced cost per base pair unit, and ease of access to technology for small research groups;
- Multiplex reactions: largest data production in a short period of time because most of the new generation equipment is exempt from individualized reactions;
- Parallelism: analysis of several individuals at the same time, allowing large population discrimination;
- Automation: minimizes human intervention, reducing operationalization errors;
- Allows work with large quantities of high accuracy data;
- Regarding platforms based on PCR and capillarity or other methods, the new technologies based on NGS show advantages as follows:
  - Sequencing from only one gene ($\sim$1 kb) to a complete genome (from many Mb to many Gb in a couple of hours);
  - Selection of several genes for expression studies, in order to analyze the expression patterns of metabolic pathways or of the complete transcriptome; and
  - Selection of a list of key genes essential to many species, and of a list of every gene in many species.

It is also emphasized that the analysis of isolated data on some genes makes less sense than checking comparative transcriptomes. Platforms currently in commercial use, despite their specificities, apply more recommended techniques for the execution of genetic analysis methodologies due to *amplicon* size, performance, and capacity of generating discriminatory sequences (Table 10.3).

For technical reasons, such as the generated amplicon size and efficacy level of equipment in generating high-quality fragments, genomic analysis methodologies based on NGSs are recommended for platforms and methodologies of sequencing according to the final results to be achieved (Table 10.4).

In Table 10.5, some particularities (e.g., the sequencing method, the kind of samples used, and the results that can be found) of each of the most used commercial platforms are presented. It is important to note that pyrosequencing produces amplicons from 400 to 1000 pb, getting very close to or exceeding automatic sequencers, which is why it is indicated for sequencing fragments demanding longer length, as is the case of full-length libraries or of complete gene sequences (Pearson and Lipman, 1988). In genome sequencing, these characteristics are desired, and even using pyrosequencing it is necessary to have a good genomic coverage. Genomic coverage greater than five times is recommended; i.e., if a gene is composed of 1.5 kb, it is necessary to sequence samples sufficiently relative to 7.5 kb in order to avoid gaps and technical errors accumulating in sequencing. The smallest the amplicon generated by sequencing,

TABLE 10.3 Use the Most Recommended Equipment for New Generation Sequencing According to the Specificity for Technology Platform

Applications	Roche GS FLX Titanium	Illumina HiSeq	ABI SOLiD
DNA sequencing	√	√	O
Resequencing	√	√	O
De novo sequencing	√	√	X
ChIP-Seq (protein/DNA)	O	√	O
Transcriptome	√	√	O
Metilome	O	√	O
Small RNA sequencing	O	√	O
Exome	O	√	O
RNASeq	O	√	O

Source: Macrogen Inc. (Seoul, Korea, 2012).√: more recommended, O: recommended, X: not recommended.

TABLE 10.4 Recommendations for Application of Sequencing Techniques Following the Methodology of Analysis of Data Required

Research	Applications	Methods
De novo sequencing	Plant whole genome	Shotgun sequencing, paired-end sequencing, de novo assembly
Resequencing	Plant whole genome	Shotgun sequencing, mapping of reference sequences
	Genomics rearrangements, number variables locus	Paired-end and mated-pair sequencing
	INDELs, SNPs, somatic mutations	Amplification for sequencing
Transcriptome analysis	Full-length transcript	cDNA transcript from mRNA or total RNA
	Multiplex paired-end sequencing tag	
	Serial Analysis Gene Expression (SAGE)	

(continued)

TABLE 10.4 Continued

Research	Applications	Methods
Gene regulation studying	Small and non-codant-RNA	
	Interaction protein/DNA (ChipSeq)	
Epigenetic changes	DNA methylation pattern	Sequencing amplicons
	Modifications of the nucleosome	Sequencing of DNA fragments
Metagenomics	Environmental interactions and DNA changes	Shotgun sequencing
	16S rRNA	Amplicons sequencing
Paleogenomics	Ancestral DNA whole genome	Shotgun sequencing
Genotyping	RNASeq	RNA sequencing
	De novo sequencing, whole genome	Shotgun sequencing, paired-end sequencing, de novo assembly
	Molecular markers screening	Probe hybridizations, oligos hybridizations, pyrosequencing, amplicons sequencing by synthesis (GBS)

the greater the genomic coverage must be. A good sequence alignment program is also required for this task. When it comes to transcripts sequencing in an Illumina platform producing short amplicons of 75 up to 100 pb, fragment libraries up to 1 kb (*paired end*), and libraries up to 5 kb (*mated pair*) are necessary. Alignment programs must be robust and genomic coverage is generally greater than 15 times. The latest technologies pledge to sequence amplicons greater than 15 kb, such as the methodology developed by Moleculo® for application in the Illumina® platform.

Researchers must choose the platform to be used for technical reasons. For instance, pyrosequencing generates a small number of long sequences, while Illumina generates a large quantity of short sequences (until now), which are limiting to the analysis and alignment in repeated regions on the genome. Most companies providing these services send aligned sequences; however, when quantification (genomic coverage or gene expression) and quality analysis of sequences is made, it is important that researchers process these data.

These technologies are widely used in the identification of SNPs, when *de novo* sequencing is made (parental genotypes sequencing) in rough form with

TABLE 10.5 Economic and Technical Specifications of the Main Platforms for Commercial Use (May 2013)

Platform	Roche 454 FLX Titanium	ABI/ Solid 3	Illumina HiSeq 2500
Sequencing methods	Sequence of synthesis: chemical pyrosequencing	Sequential ligation: cleavage of oligo sequence	Synthesis of sequence: base additional detection
Bases for run	500 m	1 to 1.5 bn	4 to 5 bn
Sample	PCR	PCR	Solid phase
Read length	100 to 400 bp	50 bp	75 to 100 bp
Run time	7.5 hr	1 to 2 days	16 to 48 hr
Cost for run	~US$10,000.00	~US$8000.00	~US$1000.00
Cost for Sanger	~US$1 bn (100×)	~US$ > 10 bn (1000×)	~US$ >90 bn (2000×)
Cost for base	~US$0.08	~US$0.005	~US$0.0010

enriched libraries or in RNASeq form (reduced genome). Implications for new techniques for the genetic improvement of plants may be:

- greater agility in obtaining molecular markers (reduction of costs, less human intervention);
- molecular marker-assisted selection with greater accuracy and speed;
- identification of large numbers of SSR and SNPs molecular markers (they are co-dominant and allow us to distinguish heterozygotes);
- use of functional molecular markers (based on genes differentially expressed) obtained from RNASeq, which have a better relationship with the phenotype;
- easier application of the technique for quantitative characteristics (high power of selecting a gene set that controls a characteristic controlled by many genes); and
- low operating cost (lower cost of labor).

RNASeq and Identification of SNPs

Using last generation equipment, such as Hiseq 2500, which generates 120 Gb in only 27 hours of work (fast mode) or 600 Gb in 10 days (slow mode), the complete sequencing of species with large genomes was overcome. However, subsequent work is still expensive and laborious. What has always been a big problem for the majority of cultivated species with large, highly redundant genomes up to 10 Gb, as in the case of sugar cane, can be easily transposed. Illumina developed a system derived from Moleculo® which produces fragments of up to 15 kb, as well as technologies based on nanopore® and PacBio. These techniques allow us to elucidate the entire gene regulation complex and its neighboring areas. Thus, total sequencing of DNA clones is an attractive alternative, being widely used in public databases, since it is fast and provides good cost/benefit ratio. RNASeq is indicated both for quantitative transcriptome analysis and SNPs identification in gene *loci*. The technique, when the Illumina HiSeq 2000 platform is used, as at March 2013, generates between 40 and 80 million sequences, which can be 100 up to 150 pb each time. Plates can be divided in {1/2}, {1/4}, and/or {1/8}. When the plate is fractioned in {1/8}, the yield is 4 up to 8 million sequences. Whereas functional genomes may contain up to 45 thousand genes, on average this represents up to 100 copies of each. If it is necessary to have greater representativeness, it is widely recommended to apply subtraction by PCR in order to enrich differentially expressed genes – which are preferential targets for genetic improvement researches – and thus these genes' representativeness becomes more than 1000 times for normally expressed genes.

Large-scale sequencing does not require (and it is not even indicated for this) the cloning of the fragments into vectors, which facilitates the preparation of samples and enables its execution. Generally, only cDNA synthesis is necessary. Initially, this level of representativeness generates a transcriptome quantitative pattern, which is extremely important to outline key genes for traits of interest using quantitative information (the number of times that the gene was sequenced is similarly proportional to the transcription level). The elevated number of copies makes it easy to obtain complete sequences of genes (when cDNA is based on *full lengths*) per alignment. These complete sequences are indicated for use in gene cloning and plant transformation.

The high number of copies is crucial to reduce sequencing technical errors aimed at the analysis and identification of SNPs. Each time, thousands of SNPs can be identified; the ability to know the characteristics and genes involved in the expression is essential to the professional in order to select them among the full or reduced transcriptome of the different genotypes (generally contrasting for the characteristics). After obtaining the SNPs, validation is one of the critical stages, which, depending on ESTs sequencing quality, can reach 50 to 90% in more rigid selection conditions, superior to those observed from those obtained in genomic regions, which are between 15 and 35%. Genotyping can be performed in a specific sequencer called GBS (genotyping by sequencing), which produces millions of reactions per day (*ultra-high throughput*), facilitating and speeding up the works.

BIOINFORMATICS TOOLS

Bioinformatics is a strategic tool for the selection of genomic regions that control plant improvement characteristics of interest. It is the science comprising all aspects of acquisition, processing, storage, distribution, analysis, and interpretation of DNA information and derivatives. It involves several mathematical and statistical tools, computing and systems biology, the objective of which is to expedite and facilitate analysis and comprehension comparing obtained × known data. Scientific effort has generated an impressive volume of molecular data from sequences of nucleotides and their likely biological functions. The challenge for researchers in dealing with the volume and complexity of data demanded greater access to analysis tools for computing. The homology of sequences allowed us to further understand genetic inheritance and its role in the expression of a characteristic.

Currently, several databases in genomics can be easily obtained through the Internet, and NCBI (www.ncbi.gov) is the most popular website. Its mission is to create automated systems for storing and analyzing information about molecular biology, biochemistry, and genetics; and to facilitate the use of these databases and research software and coordination of biotechnology information. BLAST (Altschul et al., 1990) and FASTA (Pearson, 1998) are the most used software for comparing DNA sequences with the ones from the database and they are based on the search for similarity regions.

These tools form the basis to access information on genomics since efforts should not be duplicated, avoiding ambiguity, with greater efficiency and speed in obtaining new information. To date, the integration of data storage platforms has been underutilized by researchers: in spite of the various knowledge areas (genetics, physiology, biochemistry, among others), these platforms require good computer knowledge, especially of the navigation processes and use of available bioinformatics software.

The deposition sequences in GenBank are essential to feed the genomic databases' formation system. More significant volumes of depositions were shown after 2002 (Figure 10.6). To date, the average size of the sequences stands at approximately 1 kb and it is expected that with the massive use of the

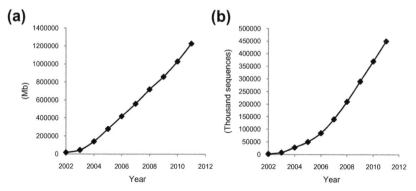

FIGURE 10.6 Deposition sequences in GenBank during the period comprising the years 2002–2011. (a) The number of bases in Mb (million base pairs) and (b) the number of sequences in the GenBank deposited. (*Source: NCBI: www.ncbi.gov*)

new generation sequencers, the number of sequences deposited will increase, but with a reduction in their size. NGS and other unprocessed sequencing data (aligned, annotated, etc.) are not deposited in the NCBI, since they have little interest in the research, besides the fact that the storage cost is currently higher than sequences generation. Often, sequencing is performed, processed, and only the result is stored "in loco," and only sequences that represent genes or their parts, or large fragments (contigs), are deposited.

In the NCBI database, 59,095,113 nucleotide sequences, as well as 71,883,697 ESTs sequences, 33,908,927 long genomic sequences, and 47,523,447 protein sequences are deposited, with approximately 450 million sequences encompassing all other classes in different organisms (available at: http://www.ncbi. nlm.nih.gov/gquery/?term=all%5Bfilter%5D, accessed February 12, 2012), in its vast majority of prokaryotes. There are a total of 6635 genome partial projects (draft), among which 1580 are from eukaryotic genomes, 9161 are from prokaryotes, and 3032 are from virus. Complete genomes comprise 2019 bacteria, 106 archaea, and 368 eukaryotes; among the last ones are found predominantly mammalian genomes and small genomes. Regarding plants, there are 5,705,806 nucleotide sequences and 23,658,239 ESTs in 410 partial and complete genomes (on February 12, 2012: http://www.ncbi.nlm.nih.gov/sites/gquery). This information constitutes an important tool for bioinformatics applied to genetic improvement, especially because the 20,130,641 SNPs and the 466,262 probes, most of them SSRs, are potential molecular markers that may be used in genotyping.

Initially, efforts were focused on the plant *Arabidopsis thaliana*, due to its small genomic size, which facilitated sequencing techniques, due to the convenience of the species cultivation. Currently, this plant's genome is very well organized and constitutes a reference for studies on other plants. Agronomic species of interest are widely represented in GenBank, with genomes of rice, wheat, sorghum, corn, barley, soybean, tomato, etc. (http://www.ncbi.nlm.nih.gov/genomes/PLANTS/PlantList.html#top).

An obstacle to genome sequencing of many plant species is their large size, although there are species with small genomes, such as *Arabidopsis thaliana*, with 157 Mb (*Genlisea margaretae* containing 63.4 Mb). However, many species of agronomic importance have very large genomes, such as sugar cane, *Saccharum hibridum*, an octaploide plant with 100–120 chromosomes distributed in eight basic genomes, containing the basic monoploid genome at a rate of 750 to 930 Mb, and the full set ranging between cultivars for approximately 10 Gb (D'Hont et al., 2008). The complete sequencing of large genomes is very easy, but the analysis, organization, and physical placement of sequences is still a very complex task, requiring large teams of researchers. Despite this barrier, genomes of many plant species have been sequenced. Due to the significant impact of the knowledge on genome sequence, completed dissection of plant biology is contained in genome sequences and their interactions. The complete sequencing (sequencing and ordering of physical map) of genomes is generally very laborious when using sequencers which produce amplicons from 400 to 600 bp; the required coverage is 8 to 12 times the genome size (if the genome size is 1 Gb, it should be sequenced from 8 to 12 Gb) when considering the high cost and intensive labor. The list of plant sequencing projects includes those that have reached the stage in which the total sequence is determined or is expected to be produced in the future. It is worth highlighting 15 projects on *Arabidopsis thaliana*, 11 on *Oryza sativa*, five on *Populus thriocarpa*, 17 on *Vitis vinifera* and one of each for *Glycine max*, *Medicago trunculata*, *Sorghum bicolor*, and *Zea mays*.

In most cases, partial sequencing (draft) is very well accepted in breeding programs, especially if it has been applied to micro-collinearity studies (gene synteny) between related species (e.g., maize and sorghum, sugar cane and sorghum, eucalyptus and poplar). Among these projects, we highlight the ones on tomato, Brachypodium, papaya, white clover, yucca, potato, and many undisclosed others.

Among the major projects in genetic mapping, studies involving large-scale sequencing (NGSS) provide a wide range of information. This requires a large number of bioinformatics researchers for data processing aimed at positioning, organizing, and delimiting the loci in their proper chromosome. In NCBI, this information is deposited in the links to the pages of MapView general overview for the specific organism. Detailed genetic mapping in plants involves 74 species with their chromosomes saturated with properly positioned loci. These loci are very important for linkage studies, especially in the application of marker-assisted selection, and in the selection phases of parents and progenies. The presence or absence of loci provides information on species or cultivar genomic structure, as well as allowing us to genotype and make predictions for parent selection.

Currently, with respect to target sequences expressed (ESTs), a large number of projects covering several plant species are gathered and at least 40,000 libraries are deposited in GenBank. For instance, sugar cane, from gender *Saccharum*, is well represented, with libraries of ESTs of 24 species in 78 projects. Of these, 55 are projects for ESTs of commercial cultivars (*Saccharum hibridum*). Among them, the Brazilian large-scale sequencing of sugarcane expressed sequence tags (ESTs) project, called SUCEST, which approached functional genome for *Saccharum hibridum* cultivar SP80-3280, of potential direct application in breeding programs. However, ESTs megaprojects for *Glycine max* comprise 1,461,624 sequences, covering 50,478 genes with 6,352,034 SNPs and 5833 probes (primers), mostly microsatellites, which makes them powerful molecular tools for immediate and future use.

The availability of these data considerably enhances the molecular biology application capacity in breeding programs assisted by molecular markers. New NGS sequencing platforms are capable of sequencing the equivalent of a eukaryotic genome in a few hours (e.g., Illumina HiSeq 2500 sequences 120 Gb in only 27 hours). This fact infers the interest in sequencing specific genotypes based on reference genome (*de novo* sequencing). Partial and transcriptome sequencing has also become more widely accessible and raises a great interest due to the possibility of connecting genotype to phenotype with an aim for greater and wider use. It is likely that RNASeq projects, NGSS platforms, and large-scale genotyping are widely available, particularly in the form of outsourced services, even facilitating access to small research groups and projects that make use of marker-assisted selection.

BIOTECHNOLOGY IN PLANT BREEDING

Among the various techniques applied to plant breeding, tissue culture plays a fundamental role. For species of asexual propagation of long life cycle or perennial, obtaining clones is essential, either by vegetative propagation (to a lesser extent), micro-propagation (medium scale), or somatic embryogenesis (large scale). The use of clones is extremely advantageous for obtaining plants with the same genetic constitution, lower phenotypic variation, and, therefore, greater homogeneity of cultivated varieties and of products obtained, which is desirable both for business and for industrial processes. Among cloning techniques, vegetative propagation by nodal segments is currently the most used due to the ease in obtaining clones and for not requiring sophisticated environmental conditions. However, the use of somatic embryogenesis tends to be spread because it is a technique that allows us to obtain thousands of clones in a small physical space, in a short period of time, and at a small cost per unit. Innovative techniques, such as the use of synthetic seeds (encapsulated somatic embryos in culture medium, ready for planting in the field) produced in automated systems, will expand the use of large-scale cloning for most cultivations of commercial interest. For the annual agronomic species, the use of double-haploid for obtaining breeding lines (discussed in Chapter 8) and protoplast fusion is potentially promising. Protoplast fusion is very important to overcome genetic barriers of extra-chromosomal inheritance, especially for angiosperms, in which mitochondrial and plastid genome are exclusively of maternal origin. The obtaining of cybrids (cytoplasmic hybrids) has been applied, for example in citrus, generating plants with higher productive capacity. The new genetic combinations of these cybrids are in the gene alleles controlling photosynthesis (located in plastid genome) and in those controlling power conversion (mitochondrial genome).

To date, genetic transformation has been widely used in basic studies of gene expression and regulation. From what is generated in scientific research, very few of the transforming events constituted commercial products, as most are intended for use in basic research. This fact will continue and even intensify, since many regulatory processes of specific genes need to be clarified. However, new guidelines will have to be evaluated in the process of plant genetic transformation, such as the use of inducible promoters, since it is necessary to control temporal and spatial transgene expression. Contrary to the constitutive promoters of CaMV 35S type, which constantly express the gene in every plant tissue, inducible and tissue-specific promoters should be prioritized in the next generation of transgenic plants. The synthetic promoters (promoters synthesized with various sequences for linkage to specific transcription factors) may also serve as a strategy in future gene constructs, in order to better control the modulation of transgene expression in plants. Still, the production of engineered synthetic genes containing deletions or insertions may increase the efficiency of transformation events. We highlight the new research involving site-directed genetic transformation, which will be able to exchange sequences in specific loci and to practice correction of defective genes or to alter protein activity. We also highlight the new classes of marker genes and selection genes, with specific sequences for their removal by effects of facilitated genetic recombination. Other processes of genetic engineering will be destined to reduce gene flow. Plastomic transformations indicate a circumstantial increase of gene expression, since we can find many plastids in a single cell, which, in theory, increases transcription level, compared to nuclear genome. It is also expected that the use of multi-transgenes stacked in a single variety, a process called pyramiding or event combination, will increase.

PERSPECTIVES OF GENOME-ASSISTED PLANT BREEDING

The new challenges of producing food for a population of over 7 billion people in a climate change environment with intensifying biotic and abiotic stresses will require genetic improvement, and the attainment of more adapted, stable plants of homogeneous production. This approach of modern cultivars should comprise the use of genomic technologies integrated with traditional techniques, in order to obtain faster results in less time and with greater magnitude. Large-scale techniques will be extremely useful for large research groups that require massive or more accurate results in a short period of time, but in the future they will also be viable for small and medium projects. However, the existence of professionals able to promote the integration of genomic techniques with conventional ones will be one of the challenges that each breeding program will not be able to neglect.

REFERENCES

Altschul, S.F., Gish, W., Miller, W., Myeres, E.W., Lipman, D.J., 1990. Basic local alignment search tool. Journal of Molecular Biology 215 (3), 403–410.

Borém, A., Teixeira, E.C., 2009. Marcadores moleculares, second ed. Suprema, Visconde do Rio Branco. 420p.

Botstein, D., White, R.L., Skolnick, M., Davis, R.W., 1980. Construction of a genetic linkage map in man using restriction fragment length polymorphisms. American Journal of Human Genetics 32, 314–331.

Cho, R., Mindrinos, M., Richards, D., Sapolsky, R., Anderson, M., Drenkard, E., Dewdney, J., Reuber, T., Stammers, M., Federspiel, N., 1999. Genome-wide mapping with biallelic markers in *Arabidopsis thaliana*. Nature Genetics 23, 203–207.

D'Hont, A., Souza, G.M., Menossi, M., Vincentz, Albert, M.G., 2008. An outstanding tropical polyploid and a major source of sucrose and bio-energy. In: Genomics of Tropical Crop Plants, first
ed. 483–513.

Hu, J., Vick, B.A., 2003. Integration of the conserved telomere sequence-derived TRAP markers onto an existing sunXower SSR linkage map. In: Plant Genome XIV, January 14–18, San
Diego, CA.

Hubby, J.L., Lewontin, R.C., 1996. A molecular approach to the study of genic heterozygosity in
natural populations I. The number of alleles at different loci in *Drosophila pseudoobscura*.
Genetics 54 (2), 577–594.

Hunter, R.L., Markert, C.L., 1957. Histochemical demonstration of enzymes separated by zone
electrophoresis in starch gels. Science 125 (3261), 1294–1295.

Jeffreys, A.J., Wilson, V., Thein, S.L., 1985. Hypervariable "minisatellite" regions of human DNA.
Nature 314, 67–73.

Konieczny, A., Ausubel, F.M., 1993. A procedure for mapping *Arabidopsis* mutations using co-
dominant ecotype-specific PCR-based makers. The Plant Journal 4 (2), 403–410.

Litt, M., Luty, J.A., 1989. A hypervariable microsatellite revealed by in vitro amplification of a
dinucleotide repeat within the cardiac muscle actin gene. American Journal of Human Genetics
44, 397–401.

Mullis, K.B., Faloona, F.A., 1987. Specific Synthesis of DNA in vitro via polymerase catalyzed
chain reaction. Methods in Enzimology 155, 335–350.

Paran, I., Michelmore, R.W., 1993. Development of reliable PCR-based markers linked to downy
mildew resistance genes in lettuce. Theoretical and Applied Genetics 85, 985–993.

Pearson, W.R., 1998. Empirical statistical estimates for sequence similarity searches. Journal of
Molecular Biology 276, 71–84.

Pearson, W.R., Lipman, D.J., 1988. Improved tools for biological sequence comparison. Proceedings of the National Academy Science USA 85 (8), 2444–2448.

Rafalski, J.A., Tingey, S.V., Williams, J.G.K., 1991. RAPD markers – a new technology for genetic
mapping and plant breeding. AgBiotech News and Information 3, 645–648.

Sax, K., 1923. The association of size differences with seed-coat pattern and pigmentation in *Phaseolus vulgaris*. Genetics 8, 552–560.

Vos, P., Hogers, R., Bleeker, M., Reijans, M., Van de Lee, T., Hornes, M., Fritjers, A., Pot, J., Pele-
man, J., Kuiper, M., et al., 1995. AFLP: a new concept for DNA fingerpinting. Nucleic Acids
Research 23, 4407–4414.

Wenzel, P., Carling, J., kudrna, D., Jaccoud, D., Huttner, E., Kleinhofs, A., Kilian, A., 2004. Diversity Arrays Technology (DArT) for whole-genome profiling of barley. PNAS New York. 101
(26), 9915–9920.

Index

A

ABRC. See Arabidopsis Biological Resource Center
Accurate phenotyping, 141
Adjacent loci, haplotype with, 87
Affymetrix® chips, 235–236
AFLPs. See Amplified fragment length polymorphism
Agrobacterium tumefaciens, 184–188
Agro-bioenergy activities, 2
Alleles, 89, 94
Allelic discrimination reactions, 36
Allelic frequencies, 88–89
Allogamous annual plant species, 130
Amiprophos-methyl (APM), 170
Amplicon, 238–240
Amplified fragment length polymorphism (AFLPs), 42
  cDNA-AFLP, 234–235
  markers, 27–30
  technique, 229
Androgenetic haploids, 209
APM. See Amiprophos-methyl
a priori distribution, 102, 117
Arabidopsis Biological Resource Center (ABRC), 154
Arabidopsis thaliana, 244–245
AroA gene, 192
Artificial microRNA, 154
Association analysis, 87
Association mapping, 93
Astrocaryum ulei, 167
Autogamous annual plant species, 130–131
Automatization, 225–226
Average binding among clusters, 55–56
Average heterozygosity, 75–76
Average-scale molecular biology techniques, 235
Averages of dissimilarity, 70
Averages of diversity, 76

B

BAC. See Bacterial Artificial Chromosome
Bacillus thuringienses, 9

Bacterial Artificial Chromosome (BAC), 148–149
Bar gene, 192
Bases for run platform, 241
BayesB method, 117
Bayesian estimation, 116–117
Bayesian methods, 124
Begomovirus genus, 198–199
Bernoulli distribution, 91
Best linear unbiased prediction (BLUP), 105–106
Bi-allelic markers, 89, 94
Biochips, 235
Bioinformatics resources, cloning of genes using, 152–154
Bioinformatics tools, 242–246
Biolistics, 188–189, 191
Biosafety of transgenic cultivars, 196–198
Biotechnology
  advances and benefits of, 7
  applications for main crop productivity and quality, 14–15
  combined classic breeding and, 12–13
Biotechnology-integrated method, 12–13
BLAST, 243
BLUP. See Best linear unbiased prediction
Bonferroni correction, 96
Breeding, 179
Bulked Segregant Analysis (BSA) technique, 144

C

Callus, 204–205
CaMV 35S type, 246–247
Candidate genes, 93
  validation of, 154–155
Capillary electrophoresis, 23–25, 29
Causal variants, 101–102
cDNA. See Complementary DNA
Cellular totipotency, 204
Central dogma of molecular biology, 227
Chemical fusion, 165
Chemical mutagens, 162

T

TAIR. See The Arabidopsis Information
    Resource
Takanotsume cultivars, 160–161
TaqMam System, 235
Target sequences expressed (ESTs), 245
    library technique, 233
T-DNA. See Transferred DNA
Thawing, 176
The Arabidopsis Information Resource
    (TAIR), 154
3' untranslated region (3' UTR), 180–181
Ti plasmid. See Tumor-inducing plasmid
Tissue culture, 8–9, 204–206
Tissue culture applications
    advantage of, 157
    for genetic improvement of plants
        embryo rescue, 166–168
        germplasm conservation and exchange,
            173–176
        mutagenesis, 162–163
        production of double haploid lines,
            168–170
        protoplast fusion, 163–166
        somaclonal variation, 157–162
        synthetic seed production, 170–172
        in vitro selection, 172–173
Tocher clustering methodology, 72
Tocher method, 71
Total genetic heterozygosity, 76
Total identity of subpopulations, 75
Transcription, 231–232
Transcriptome analysis research,
    239–240
Transcriptome-based methods, 12
Transcriptome sequencing, 245–246
Transferred DNA (T-DNA), 184–185
Transfer RNA (tRNA), 180
Transformation, 182–183
Transformed cells, regenerating plants from,
    192
Transgene-derived RNA, 198–199
Transgenic cultivars
    biosafety of, 196–198
    commercial, 198–199
    use, effects, and management of, 193–199
Transgenic plants
    development
        laboratory steps for, 190–192

methodologies for, 184–190
identification of, 192–193
Transient expression, 154–155
TRAP technique, 230
Triparental conjugation, 186–187
tRNA. See Transfer RNA
Tumefaciens, 184
Tumor-inducing (Ti) plasmid, 184, 186
    of *Agrobacterium tumefaciens*, 185
Tungsten microparticles in transformation
    process, 188
Turbinicarpus, 174–175
Two-factor model classification, 79

U

Unweighted index, 52

V

Value for cultivation and use (VCU)
    testing, 12
Values of intercluster distances, 72
Variance analysis, 79
Variance components
    estimation of, 79–80
    and statistics, statistic tests of, 80
        association between, 80
VCU testing. See Value for cultivation and use
    testing
Vegetative propagation, perennial plant species
    and species with, 130
Virulence genes, 184

W

Ward's minimum variance method, 57
Weighted average binding among clusters,
    56–58
Weighted index, 52
Wide crosses, 206–208
Woody plant medium (WPM), 163
WPM. See Woody plant medium

X

χ^2 Test, probability level for, 137–139

FIGURE 2.4 Polymorphism detection in a 6% polyacrylamide denaturing gel stained with silver nitrate. Genotypic mapping of a *Coffea canephora* population resulting from the crossing of two heterozygous parents (P1 × P2).

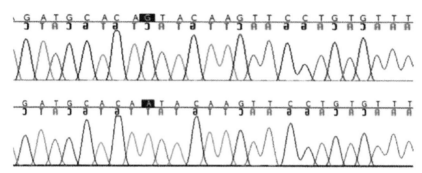

FIGURE 2.12 Identification of SNPs in different individuals by alignment of DNA sequences. The SNP in the example corresponds to the replacement of G with A.

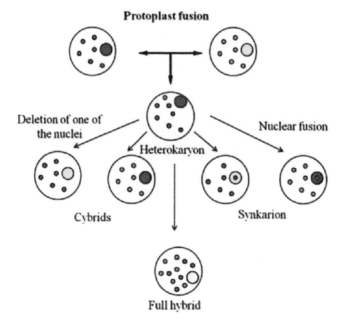

FIGURE 7.2 Diagram of full hybrid formation from protoplast fusion.

FIGURE 7.4 Development of encapsulated explants of *Corymbia torelliana* × *C. citriodora* under aseptic and non aseptic conditions. (a) Capsules formed by apex (a1) or nodal segment (a2) encapsulation. (b) Regenerated sprouts derived from capsules after 4 weeks in different culture media (b1, 1/2 MS; b2, MS; b3, MS + 2.2 μM BA; and b4, MS + 2.2 μM BA + 0.3 μM NAA). (c) Roots (arrows) induced in synthetic seeds. (d) Seedlings formed directly from sprouts after pretreatment with AIB at different levels (d1, 0; d2, 4.9; d3, 19.6; or d4, 78.4 μM). (e) Potted seedlings after 2 weeks of acclimatization. (f) Synthetic seeds after 4 weeks of pre conversion. (g) Seedlings formed directly from synthetic seeds after pre conversion and planting in different non sterile substrates for 4 weeks (g1, vermiculite and perlite; g2, substrates; g3, organic compound). (h) Potted seedlings at 2 months of age. Scale bar = 1 cm. (*Adapted from Hung and Trueman, 2012*)

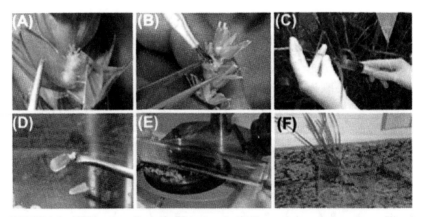

FIGURE 9.5 Wide crosses (interspecific crosses). (A) Emasculation of wheat plants; (B) pollination with maize pollen; (C) application of 2,4-D and AgNO₃ solution; (D) embryo rescue; (E) transfer to tissue culture; and (F) double-haploid plant produced.

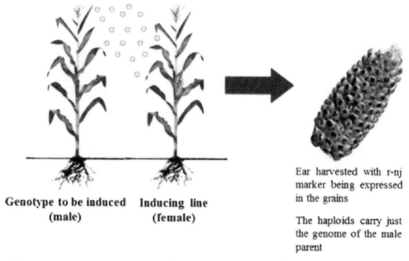

Genotype to be induced   Inducing line
(male)   (female)

Ear harvested with r-nj
marker being expressed
in the grains

The haploids carry just
the genome of the male
parent

FIGURE 9.8 Haploid production through androgenetic inducing lines. (*Source: IAPAR*)

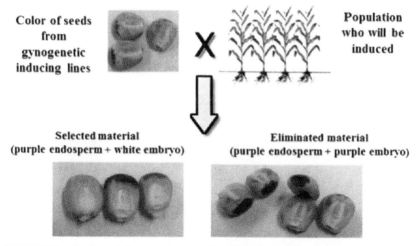

Color of seeds
from
gynogenetic
inducing lines

X

Population
who will be
induced

Selected material
(purple endosperm + white embryo)

Eliminated material
(purple endosperm + purple embryo)

FIGURE 9.9 Morphological marker for seed color to identify haploids. (*Source: IAPAR*)

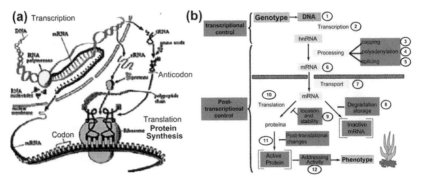

FIGURE 10.1 **Detailing from the central dogma of the molecular biology.** (a) Process for transcription of DNA into RNA and translation of mRNA into protein. (b) Control of gene expression in eukaryotes: (1) epigenetic control: methylation, acetylation, and phosphorylation of histones and other proteins and DNA, (2) control at the transcriptional level: changes in assembly conditions, processivity and/or stop the RNA polymerase complex, (3) capping: changes in the activity of the complex 7-methyl-guanylyl-transferase on the 5′ end of hnRNA, (4) polyadenylation: activity and polyadenylation complex formation and activation of polyA-polymerase, (5) splicing: timely correct assembly of the spliceosome snRNP and the removal of introns, (6) addressing the mRNA: RNA binding protein to protect and mRNA export, (7) transport: addressing and protection of mRNA to ribosomes in the cytosol, (8) degradation or storage: interaction with small RNA followed by degradation or folding temporary, (9) location and stability: correct target mRNA and suitable environmental conditions for the cytosol, (10) translation: assembly of complex translation (initiation and elongation), (11) post-translational changes: folding, phosphorylation, oxidation, and protein sulfurization, (12) addressing and activity of protein: protein binding addressing, encapsulation, glycosylation, oxidation, and interacting protein/protein, (13) regulating in the substrate level: cofactors, substrates, and products (metabolism), and functional activity of the protein or enzyme.

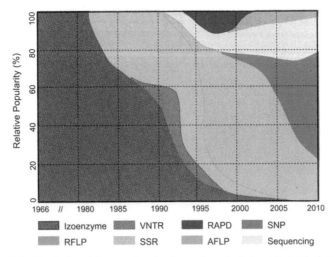

FIGURE 10.3 Popularity of the use of molecular markers (including biochemical marker isoenzyme) during the period from 1966 to 2010, according to publications in international journals.

Printed and bound by CPI Group (UK) Ltd, Croydon, CR0 4YY

03/10/2024

01040419-0002